Adarsh Anand and Mangey Ram (Eds.)
**Systems Performance Modeling**

# De Gruyter Series on the Applications of Mathematics in Engineering and Information Sciences

Edited by
Mangey Ram

## Volume 4

# Systems Performance Modeling

Edited by
Adarsh Anand and Mangey Ram

DE GRUYTER

**Editors**
Adarsh Anand
Faculty of Mathematical Sciences
University of Delhi
110007 Delhi, India
adarsh.anand86@gmail.com

Mangey Ram
Department of Computer Science and Engineering
Graphic Era Deemed to be University
566/6 Bell Road
248002 Clement Town, Dehradun, Uttarakhand, India
drmrswami@yahoo.com

ISBN 978-3-11-060450-4
e-ISBN (PDF) 978-3-11-061905-8
e-ISBN (EPUB) 978-3-11-060763-5
ISSN 2626-5427

**Library of Congress Control Number: 2020948102**

**Bibliographic information published by the Deutsche Nationalbibliothek**
The Deutsche Nationalbibliothek lists this publication in the Deutsche Nationalbibliografie; detailed bibliographic data are available on the Internet at http://dnb.dnb.de.


Cover image: MF3d/E+/Getty Images
Typesetting: Integra Software Services Pvt. Ltd.
Printing and binding: CPI books GmbH, Leck

www.degruyter.com

MIX
Papier aus verantwortungsvollen Quellen
FSC www.fsc.org
FSC® C083411

# Acknowledgment

The editors acknowledge Walter de Gruyter and the editorial team for their adequate and professional support during the preparation of this book. They thank all authors who have contributed to this editorial work. Reviewers, who have helped through their comments and suggestions in improving the quality of the chapters, deserve significant praise for their assistance.

Finally and most importantly, editors dedicate this editorial book to their family and friends.

Adarsh Anand
University of Delhi, India

Mangey Ram
Graphic Era Deemed to be University, India

https://doi.org/10.1515/9783110619058-202

# Preface

Virtually all countries now depend on complex computer-based systems. Infrastructure and utilities rely on the computer-based systems, and almost everything that surrounds us includes a computer and controlling software. Information and communication technologies have been at the heart of economic changes for more than a decade, notably in all areas of businesses. In line with this, this edited issue of *System Performance Modeling* includes invited papers that deal with the modeling and analysis of software systems with the aid of computers. Emphasis is on concepts, theories, and techniques developed in the infocom discipline. The topics covered are organized as follows:

Chapter 1 discusses software vulnerability patch management, which is one of the newer areas of research.

Chapter 2 discusses the debugging process for modeling quality and reliability aspects of the software project management.

In Chapter 3, an analysis pertaining to vehicular cloud computing has been presented.

Chapter 4 presents a comparative study dealing with the agile methodology inculcating increasing failure rate software reliability models.

In today's time, everyone is talking and dealing with the three Vs, which we know by the name of big data. Chapter 5 presents a mathematical framework to model the fault big data analysis based on effort estimation and artificial intelligence for Open Source Software Project.

Every system, be it hardware or software, requires data. The more accurate the data, the more are the chances that the results will be error-free. Chapter 6 presents a modeling framework that deals with the meaning of data streams and its impact on the system performance.

Chapter 7 discusses assessing the reliability of public switched telephone network, which is the most useful telecommunication network in general.

No software is complete unless some articulation is talked about for hardware, as both go hand in hand when talked about a complex system. Chapter 8 contains a description of the utility of Weibull failure laws for reliability measures of a series–parallel system.

Adarsh Anand
University of Delhi, India

Mangey Ram
Graphic Era Deemed to be University, India

https://doi.org/10.1515/9783110619058-203

# About the editors

**Adarsh Anand** did his doctorate in the area of Software Reliability Assessment and Innovation Diffusion Modeling in Marketing. Presently, he works as an assistant professor in the Department of Operational Research, University of Delhi (INDIA). He has been conferred with Young Promising Researcher in the field of Technology Management and Software Reliability by Society for Reliability Engineering, Quality and Operations Management (SREQOM) in 2012. He is a lifetime member of the Society for Reliability Engineering, Quality and Operations Management (SREQOM). He is also on the editorial board of *International Journal of System Assurance and Engineering Management* (Springer). He has guest edited several special issues for journals of international repute. He has publications in journals of national and international repute. His research interest includes software reliability growth modeling, modeling innovation adoption and successive generations in marketing, and social network analysis. He has worked with CRC Press for two editorial projects; "System Reliability Management: Solutions and Technologies" and "Recent Advancements in Software Reliability Assurance." He has also authored one textbook with CRC group "Market Assessment with OR Applications."

**Dr. Mangey Ram** received the Ph.D. degree major in mathematics and minor in computer science from G. B. Pant University of Agriculture and Technology, Pantnagar, India. He has been a faculty member for around 12 years and has taught several core courses in pure and applied mathematics at undergraduate, postgraduate, and doctorate levels. He is currently a professor at Graphic Era (Deemed to be University), Dehradun, India. Before joining the Graphic Era, he was a deputy manager (probationary officer) with Syndicate Bank for a short period. He is the editor in chief of *International Journal of Mathematical, Engineering and Management Sciences*, and the guest editor and member of the editorial board of various journals. He is a regular reviewer for international journals, including IEEE, Elsevier, Springer, Emerald, John Wiley, Taylor & Francis, and many other publishers. He has published 175 plus research publications in IEEE, Taylor & Francis, Springer, Elsevier, Emerald, World Scientific, and many other national and international journals of repute and also presented his works at national and international conferences. His fields of research are reliability theory and applied mathematics. Dr. Ram is a senior member of the IEEE, life member of Operational Research Society of India, Society for Reliability Engineering, Quality and Operations Management in India, Indian Society of Industrial and Applied Mathematics, member of International Association of Engineers in Hong Kong, and Emerald Literati Network in the UK. He has been a member of the organizing committee of a number of international and national conferences, seminars, and workshops. He has been conferred with the *Young Scientist Award* by the Uttarakhand State Council for Science and Technology, Dehradun, in 2009. He has been awarded the *Best Faculty Award* in 2011; *Research Excellence Award* in 2015; and recently, *Outstanding Researcher Award* in 2018 for his significant contribution in academics and research at Graphic Era Deemed to be University, Dehradun, India.

https://doi.org/10.1515/9783110619058-204

# Contents

# List of contributing authors

**Adarsh Anand**
Department of Operational Research
University of Delhi
Delhi, India
adarsh.anand86@gmail.com

**Nestor Ruben Barraza**
Universidad Nacional de Tres de Febrero
Buenos Aires, Argentina
nbarraza@untref.edu.ar

**S. K. Chauhan**
Shaheed Rajguru College of Applied Sciences for Women
University of Delhi
New Delhi, India
statskumar1@gmail.com

**Mario Diván**
National University of La Pampa
La Pampa, Argentina
mjdivan@eco.unlpam.edu.ar

**Anu A. Gokhale**
Illinois State University
Normal, Illinois, USA
aagokhale@ilstu.edu

**Shivani Gupta**
University of Delhi
Delhi, India
shivani222gupta@gmail.com

**Vandana Gupta**
University of Delhi
Delhi, India
khaitan.vandana@gmail.com

**Shinji Inoue**
Kansai University
Osaka, Japan
ino@kansai-u.ac.jp

**Jasmine Kaur**
University of Delhi
Delhi, India
jasminekaur.du.aor@gmail.com

**S. C. Malik**
Maharshi Dayanand University
Rohtak, India
sc_malik@rediffmail.com

**Kuldeep Nagiya**
Graphic Era Deemed to be University
Dehradun, Uttarakhand, India
kuldeepnagiya@gmail.com

**Gabriel Ricardo Pena**
Universidad Nacional de Tres de Febrero
Buenos Aires, Argentina
gpena@untref.edu.ar

**Mangey Ram**
Department of Mathematics, Computer Science and Engineering
Graphic Era Deemed to be University
Dehradun, Uttarakhand, India
drmrswami@yahoo.com

**María Laura Sánchez Reynoso**
National University of La Pampa
La Pampa, Argentina
mlsanchezreynoso@eco.unlpam.edu.ar

**Yoshinobu Tamura**
Tokyo City University
Tokyo, Japan
tamuray@tcu.ac.jp

**Shigeru Yamada**
Tottori University
Tottori, Japan
yamada@tottori-u.ac.jp

https://doi.org/10.1515/9783110619058-206

Adarsh Anand, Jasmine Kaur, Anu A. Gokhale, and Mangey Ram

# 1 Impact of available resources on software patch management

**Abstract:** Software security has been an area of immense research as most of the things surrounding us are technology based. Much has been talked about vulnerabilities, their categories and types. Some studies elaborated and extended the available discovery models but few have considered the correction process in the same work. In this study, an approach to deal with software vulnerability through the release of patch/updates has been presented. The methodical work presented here discusses a mathematical model for optimal allocation of resources to remove vulnerabilities through an update.

**Keywords:** severity, software patch, software security, software updates, vulnerability, vulnerability correction model

## 1.1 Introduction

Software security is a matter of grave concern, and the need for secure software has been stressed upon enough. The first half of the year 2019 had already seen 3,800 cases of publicly disclosed software breaches and 4.1 billion records exposed [1]. With the hacking tools getting more and more advanced, it is a tough fight to keep the system safe from the intruders. Even minor loopholes or oversights leave the software highly vulnerable. Often, the exploited loopholes have a fix available, but due to lack of comprehension of the situation's gravity, they are never plugged. Such faults or glitches in the system architecture, design, code or implementation that compromises the software's security are termed as software vulnerabilities. The most common vulnerabilities are caused by memory safety violations, input validation error, privilege confusion bugs, privilege escalation, race conditions, side channel attack, and user interface failure. The most common types of vulnerabilities are SQL injection, command injection, buffer overflow, uncontrolled format string, integer overflow, cross-site scripting, and so on. The software vulnerabilities are also categorized on the basis of the extent of damage their presence or exploitability causes. The Common Vulnerabilities and Exposures (CVE) database provides a Common Vulnerability Scoring System (CVSS) score to the reported vulnerabilities known as the severity

**Adarsh Anand, Jasmine Kaur,** University of Delhi, Delhi, India
**Anu A. Gokhale,** Illinois State University, USA
**Mangey Ram,** Graphic Era Deemed to be University Dehradun, Uttarakhand, India

https://doi.org/10.1515/9783110619058-001

index on an ascending scale of 1 to 10 [2]. On the basis of severity score, vulnerabilities are categorized as low, medium, high, and critical. The vulnerability discovery process is modeled through Vulnerability Discovery Models (VDMs), which help quantify the vulnerabilities discovered and understand their detection behavior with time. The foremost software VDM was the Anderson's thermodynamic model [3]. Rescorla [4] presented a linear and exponential trend in the vulnerability detection process. The Alhazmi-Malaiya model, an S-shaped, logistic VDM defined the three phases in the vulnerability discovery process as linear, learning, and saturation [5]. An effort-based model was proposed by Alhazmi and Malaiya [6] to model the effort consumed in terms of resources and budget in finding the vulnerabilities. Arora et al. [7] had discussed the optimal policy for vulnerability disclosure. Kapur et al. [8] proposed a logistic rate in the vulnerability detection model. The model by Anand and Bhatt [9] proposed a hump-shaped vulnerability detection rate. Bhatt et al. [10] discussed a VDM that categorizes the discovered vulnerabilities as leading vulnerabilities and additional vulnerabilities. Vulnerability discovery process over multiple versions of the software has been modeled by Anand et al. [11].

The software firms tend to test the software for vulnerabilities and release a remedial patch, or a bunch of patches simultaneously known as updates, depending on the number of vulnerabilities to be catered. Software patches are sets of corrective code meant to replace the defective, fault-causing code and thus prevent exploitability. Beattie et al. [12] presented a mathematical model to determine the best time to apply a security patch for software vulnerability. The utility of patches in improving the software was first mathematically discussed in a Software Reliability Growth Model (SRGM) by Jiang and Sarkar [13], Arora et al. [14] had discussed the impact of patching on the software quality. Das et al. and Deepika et al. [15, 16] had explored the role of the tester and the user in software reliability growth via the patch service. Anand et al. [17] highlighted the importance of providing patching service in the software and proposed a scheduling policy for the optimal release of software. Anand et al. [18] considered faults and vulnerabilities simultaneously and presented optimal patch release policy. Recently, Kaur et al. [19] presented the first vulnerability correction model (VCM) and discussed the mathematical frame for the time gap between the correction of leading and dependent vulnerabilities.

Precautions need to be taken during the software development process to minimize the possibility of vulnerabilities. Better coding practices, better organizational practices, exhaustive testing, and so on, can help reduce the number of vulnerabilities in the software. The presence of vulnerability itself in the software is not the main concern but rather their exploitability is. It is not possible to ensure vulnerability free software but their number can be minimized so as to avoid damage through exploitability. After the release of the software the only way to ensure software security is continuous and exhaustive testing of the software. The developmental resources in the form of manpower, budget, time, and so on, available throughout the project are in a limited supply. The available resources are allocated to each phase of the Software Development Life Cycle (SDLC) and a major portion of it goes in software testing (both before and after software release). We know that testing the software in-house is cheaper as compared to field testing, thus release of patches and updates as part of maintenance activities is expensive. During the vulnerability removal process, usually preference is given to vulnerabilities with higher severity as they are likely to cause bigger damage. But the less severe vulnerabilities cannot be ignored as they too are dangerous and can cause critical damage upon exploitation. The number of vulnerabilities that can be dealt in an update depends on many factors such as the number of vulnerabilities in the software, the resources available to patch them, and the nature/severity of the vulnerability. Hence, an optimal allocation of resources is very essential for the operational phase. A trade-off needs to be understood to plan an update so that the maximum number of vulnerabilities can be catered with the proper utilization of the limited resources at hand. Since the number of vulnerabilities in software is usually large, not all vulnerabilities can be handled in a single update. For a singular issue or for dealing with a handful of vulnerabilities, patches are released. But when a large number of vulnerabilities need to be dealt with, the patches are clubbed together and released to the user as an update. For a particular version of the software, multiple singular patches or updates can be released depending on the need of the software.

In literature, the resource allocation problem has been extensively explored in the context of software reliability growth models [20–30]. A lot has been discussed and presented in past in the context of SRGMs but there is limited work done when VDM is talked about from resource allocation point of view. A recent work by Bhatt et al. [31] is the first to discuss an optimal resource allocation model for the vulnerability discovery process.

The optimization problem to maximize a quantity (vulnerability correction) while trying to minimize the use of the other (resource utilization), has been taken up in the current work. An effort has been made to allocate an optimal amount of resources for software vulnerability correction process. The allocation of the resources will lead to increased security, improved reliability, and quality of the software. The flow of the model can be understood through Figure 1.1.

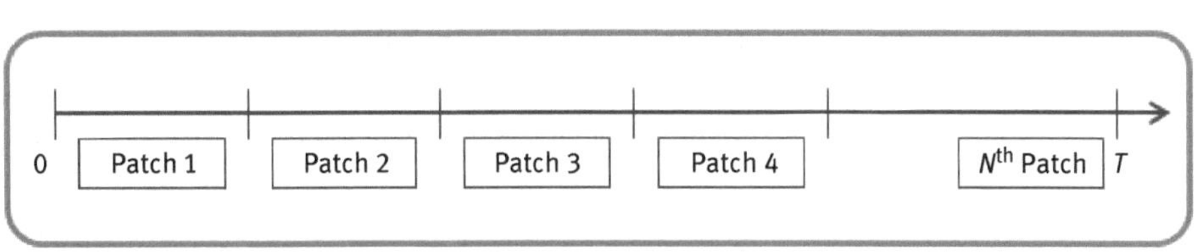


**Figure 1.1:** Flow of the model.

The chapter has been bifurcated as follows: Section 1.2 discusses the notations and model development, Section 1.3 presents a numerical example to validate the proposed model, and Section 1.4 concludes the chapter and is followed by a list of the references used.

## 1.2 Model development

### 1.2.1 Notations

The notations used in the model development are:

$M$ : Number of severity level groups pertaining to software vulnerability

$i$ : Variable representing the severity level group, $i = 1, 2, \ldots, M$

$N_{Ri}$ : Expected number of vulnerabilities of $i$th severity group

$r_{Ri}$ : Vulnerability removal rate

$y_i(t)$: Resource utilization at time $t$ for $i$th severity group and $Y_i(t) = \int_0^t y_i(w)dw$ for $i$th severity group

$Y_i^*$ : Optimal value of $Y_i$, $i = 1, 2, \ldots, M$

$Z$ : Total resources available

$\Omega_{Ri}(t)$ : Number of vulnerabilities removed in $(0, t]$ of the $i$th severity group, that is, mean value function of the Non Homogeneous Poisson Process (NHPP)

$\Omega_{Ri}(Y_i(t))$: Cumulative number of vulnerabilities removed using resources $Y_i(t)$ in time $(0, t]$

$T$ : Total time available for the vulnerability removal process

### 1.2.2 Resource allocation problem

The work in vulnerability modeling extensively deals with either the vulnerability detection process or the vulnerability exploitation process. The vulnerability correction process has not obtained much attention even though it is a very important aspect of the vulnerability life cycle. The VCM presented by Kaur et al. [19] describes the vulnerability removal/fixation phenomenon for the detected vulnerabilities. They had further categorized the removed vulnerabilities as leading and dependent. In this section, we shall extend upon their work and present another VCM that will then be used to allocate resources for releasing updates. The work of Rescorla [4] along with the proposals of Alhazmi and Malaiya [6] has been considered to design our present mathematical structure.

Recently, Bhatt et al. [31] had presented a model to dynamically allocate resources to discover vulnerabilities of varying severity levels. The following differential equation defines the relation between the effort consumed in detecting vulnerabilities of $i$th severity level and the number of vulnerabilities discovered:

$$\frac{d\Omega_i(t)}{dt} \Big/ x_i(t) = r_i(N_i - \Omega_i(t)), \quad i = 1, 2, \ldots, n, \tag{1.1}$$

where $\Omega_i(t)$denotes the number of vulnerabilities of $i$th severity detected till time $t$; $N_i$ is the number of vulnerabilities detected of $i$th severity and $r_i$ represents their detection rate, while $x_i(t)$ represents the resources or effort spent in vulnerability detection.

Solution of the eq. (1.1) gives

$$\Omega_i(t) = N_i\left(1 - e^{-r_i.X_i(t)}\right), \qquad i = 1, 2, \ldots, n, \tag{1.2}$$

Using the above-mentioned analogy, the effort consumption in the vulnerability removal phenomena can be defined on the lines of effort-based vulnerability detection phenomena. According to the work of Kaur et al. [19], the VCM is an NHPP and effort-based VDM is NHPP; hence, the effort-based VCM would also follow NHPP.

The VCM can thus be defined as

$$\frac{d\Omega_{Ri}(t)}{dt} \Big/ y_i(t) = r_{Ri}(N_{Ri} - \Omega_{Ri}(t)), \quad i = 1, 2, \ldots, M. \tag{1.3}$$

Solving the differential equation with initial conditions $t = 0$, $Y_i(t) = 0$ and $\Omega_{Ri}(t) = 0$, we obtain the relation as

$$\Omega_{Ri}(t) = N_{Ri}\left(1 - e^{-r_{Ri}.Y_i(t)}\right), \qquad i = 1, 2, \ldots, M. \tag{1.4}$$

Using the exponential distribution to define the behavior of the effort function as described by Bhatt et al. [31], we obtain

$$\frac{dY_i(t)}{dt} = g_i(\lambda_i - Y_i(t)), \qquad i = 1, 2, \ldots, M, \tag{1.5}$$

where $g_i$ represents the rate at which the available resources are consumed in the vulnerability removal of the $i$th severity level group and $\lambda_i$ denotes the total resources that are available to remove vulnerabilities of a given severity level group. The solution of the above-mentioned differential equation gives us

$$Y_i(t) = \lambda_i\left(1 - e^{-g_i.t}\right), \quad i = 1, 2, \ldots, M. \tag{1.6}$$

The purpose of our optimization problem is to remove the maximum number of vulnerabilities of varying severity levels in a given update keeping in mind the limited resources available. Hence, an optimization problem can be formulated as:

$$\begin{aligned} &\text{Maximize} \sum_{i=1}^{M} \Omega_{Ri}(Y_i) = \sum_{i=1}^{M} N_{Ri}(1 - e^{-r_{Ri}Y_i}) \\ &\text{subject to} \sum_{i=1}^{M} Y_i \le Z,\, Y_i \ge 0, \quad \text{where } i = 1, \ldots, M. \end{aligned} \tag{1.7}$$

The solution of the optimization problem allocates resources in a particular update, that is, it helps the debugging team allocate its limited resources based on the number and nature and of the vulnerability. The leftover vulnerabilities will be catered to in the next update via a software patch. A run-through of the model has been shown via a numerical illustration in the following section.

## 1.3 Numerical illustration

For the purpose of model validation a simulated dataset has been considered, which contains the vulnerability removal data. The dataset contains 636 vulnerabilities that are further categorized on the basis of their severity levels. Six severity level groups have been designed as shown in the first column of Table 1.1. It has been assumed that we have a fixed supply of resources of 3,100 units. But we have not allocated all our resources in a single go and wish to divide it proportionately between the multiples updates as discussed earlier. For the first update, we have limited ourselves to 1,000 units of resources. We have used the above-discussed VCM (eqs. (1.4) and (1.6)) to estimate the value of $r_R$ and allocate these 1,000 units of resources for the different severity groups. Software packages SPSS and LINGO have been utilized for parameter estimation and for solving the optimization problem respectively. Table 1.1 represents the dynamically allocated resources ($Y$) for each severity group, number of vulnerabilities removed ($\Omega_R$) on the basis of the allocated resources.

As given in Table 1.1, approximately 72% of the initial 636 vulnerabilities have been removed in the first update itself. This amount contains a propionate removal from each severity group based on their initial content. This high percentage of removal has been achieved due to various factors such as rigorous debugging (both due to in-house and field testing) and user participation. It is important to remove the major portion of the vulnerabilities in the first update itself so that the reliability can be maximized as soon as possible.

**Table 1.1:** Resource allocations for update 1.

| Severity groups | $N_R$ | $r_R$ | $Y$ | $\Omega_R$ | Percentage of vulnerabilities removed | Percentage of vulnerabilities remaining | Number of leftover vulnerabilities |
|---|---|---|---|---|---|---|---|
| 1–3 | 203.79 | 0.0051 | 303.1525 | 160.0106 | 78.5167 | 21.4833 | 43.7813 |
| 4 | 65.86 | 0.0080 | 107.8778 | 37.9292 | 57.5879 | 42.4121 | 27.9339 |
| 5 | 72.92 | 0.0122 | 113.6826 | 54.7453 | 75.0782 | 24.9218 | 18.1724 |
| 6–7 | 34.95 | 0.0290 | 52.3435 | 27.2873 | 78.0842 | 21.9158 | 7.6587 |
| 8–9 | 205.41 | 0.0035 | 335.1865 | 142.4713 | 69.3603 | 30.6397 | 62.9363 |
| 10 | 53.22 | 0.0125 | 87.7571 | 35.4596 | 66.6292 | 33.3708 | 17.7597 |
| **Total** | **636.15** | – | **1,000.0000** | **457.9032** | **71.9809** | **28.0191** | **178.2423** |

In the second update the leftover vulnerabilities from the first update will be catered to. We have allotted 800 units of available resources to this update. Following the approach discussed earlier, we run the second iteration to remove the leftover vulnerabilities.

Table 1.2 shows that out of the approximately 178 vulnerabilities to be dealt in this update, 63% of them have been removed and approximately 66 vulnerabilities are still latent in the system. Thus, the debugging process continues and the need of another patch comes into picture. The third update has been allotted 700 units of resources to remove the 65 vulnerabilities.

By the third update a very large portion of the initial vulnerabilities has been removed and the software will be perceived to be quite reliable. As can be seen in Table 1.3, approximately 58% of the vulnerabilities of the third update have been removed. Now we are left with only 28 vulnerabilities out of the initial 636. So we allocate our remaining 600 units of resources to remove these vulnerabilities.

Table 1.4 shows that with the given amount of resources we have been able to remove 50% of the vulnerabilities of the fourth update. By the time the fourth update reaches the user, the useful life of the version is almost over and further resources will not be spent on it. If the leftover vulnerabilities are crucial and removing them is of utmost importance, then the firm will have to allocate more resources. If the initial count of the vulnerabilities in the system is compared to the number of vulnerabilities still latent after the last update, we can see that approximately 98% of the vulnerability content could be eliminated using our approach. Thus, the software can be considered to be highly secure. Further, our approach has given equal importance to vulnerabilities of all severity groups and not just focused on the high severity groups while ignoring the low severity vulnerabilities. An implicit result that can also be drawn from the above-mentioned numerical

**Table 1.2:** Resource allocations for update 2.

| Severity groups | $N_R$ | $r_R$ | $Y$ | $\Omega_R$ | Percentage of vulnerabilities removed | Percentage of vulnerabilities remaining | Number of leftover vulnerabilities |
|---|---|---|---|---|---|---|---|
| 1–3 | 43.78 | 0.0051 | 196.5014 | 27.6243 | 63.0961 | 36.9039 | 16.1570 |
| 4 | 27.93 | 0.0080 | 125.3756 | 17.6253 | 63.0964 | 36.9036 | 10.3086 |
| 5 | 18.17 | 0.0122 | 81.5610 | 11.4660 | 63.0956 | 36.9044 | 6.7064 |
| 6–7 | 7.66 | 0.0290 | 34.3757 | 4.8325 | 63.0977 | 36.9023 | 2.8262 |
| 8–9 | 62.94 | 0.0035 | 282.4744 | 39.7103 | 63.0961 | 36.9039 | 23.2260 |
| 10 | 17.76 | 0.0125 | 79.7119 | 11.2058 | 63.0970 | 36.9030 | 6.5539 |
| **Total** | **178.24** | **–** | **800.0000** | **112.4641** | **63.0962** | **36.9038** | **65.7781** |

**Table 1.3:** Resource allocations for update 3.

| Severity groups | $N_R$ | $r_R$ | $Y$ | $\Omega_R$ | Percentage of vulnerabilities removed | Percentage of vulnerabilities remaining | Number of leftover vulnerabilities |
|---|---|---|---|---|---|---|---|
| 1–3 | 16.16 | 0.0051 | 171.9406 | 9.4033 | 58.1992 | 41.8008 | 6.7538 |
| 4 | 10.31 | 0.0080 | 109.7074 | 5.9997 | 58.2004 | 41.7996 | 4.3090 |
| 5 | 6.71 | 0.0122 | 71.3639 | 3.9029 | 58.1973 | 41.8027 | 2.8035 |
| 6–7 | 2.83 | 0.0290 | 30.0734 | 1.6447 | 58.1939 | 41.8061 | 1.1815 |
| 8–9 | 23.23 | 0.0035 | 247.1680 | 13.5173 | 58.1992 | 41.8008 | 9.7086 |
| 10 | 6.55 | 0.0125 | 69.7468 | 3.8143 | 58.1991 | 41.8009 | 2.7396 |
| **Total** | **65.78** | **–** | **700.0000** | **38.2822** | **58.1990** | **41.8010** | **27.4959** |

illustration is that, based on a given resource pool and the number of leftover vulnerabilities, we can determine the number of patches/updates that will be required to deal with all the vulnerabilities of a particular version of the software.

**Table 1.4:** Resource allocations for update 4.

| Severity groups | $N_R$ | $r_R$ | $Y$ | $\Omega_R$ | Percentage of vulnerabilities removed | Percentage of vulnerabilities remaining | Number of leftover vulnerabilities |
|---|---|---|---|---|---|---|---|
| 1–3 | 6.75 | 0.0051 | 147.3758 | 3.5560 | 52.6516 | 47.3484 | 3.1978 |
| 4 | 4.31 | 0.0080 | 94.0226 | 2.2686 | 52.6486 | 47.3514 | 2.0404 |
| 5 | 2.80 | 0.0122 | 61.1601 | 1.4759 | 52.6450 | 47.3550 | 1.3276 |
| 6–7 | 1.18 | 0.0290 | 25.7957 | 0.6223 | 52.6723 | 47.3277 | 0.5592 |
| 8–9 | 9.71 | 0.0035 | 211.8555 | 5.1118 | 52.6516 | 47.3484 | 4.5969 |
| 10 | 2.74 | 0.0125 | 59.7903 | 1.4426 | 52.6563 | 47.3437 | 1.2970 |
| **Total** | **27.50** | – | **599.9999** | **14.4771** | **52.6518** | **47.3482** | **13.0188** |

## 1.4 Conclusion

The work presents a vulnerability removal model. An optimization model has been utilized to allocate the resources to remove the vulnerabilities of different severity that is achieved via multiple patch/updates. The proposed model has been validated on simulated vulnerability data. Approximately 98 percent of the vulnerabilities were removed through the proposed model. Thus, with the help of the current work, the resource allocation for vulnerability removal can be optimally achieved.

## Bibliography

[1] https://us.norton.com/internetsecurity-emerging-threats-2019-data-breaches.html
[2] https://www.cvedetails.com
[3] Anderson, R. (2002). Security in open versus closed systems – the dance of Boltzmann, Coase and Moore, Technical report, Cambridge University, England.
[4] Rescorla, E. (2005). Is finding security holes a good idea?, IEEE Security & Privacy, 3(1), 14–19.
[5] Alhazmi, O.H. and Malaiya, Y.K. (2005a). Quantitative vulnerability assessment of systems software, In proceedings of Annual Reliability and Maintainability Symposium, IEEE, 615–620.
[6] Alhazmi, O.H. and Malaiya, Y.K. (2005b). Modeling the vulnerability discovery process, In 16th IEEE International Symposium on Software Reliability Engineering (ISSRE'05), IEEE, Chicago, IL, 138–147.
[7] Arora, A., Telang, R. and Xu, H. (2008). Optimal policy for software vulnerability disclosure, Management Science, 54(4), 642–656.

[8] Kapur, P.K., Sachdeva, N. and Khatri, S.K. (2015). Vulnerability Discovery Modeling, International Conference on Quality, Reliability, Infocom Technology and Industrial Technology Management, 34–54.
[9] Anand, A. and Bhatt, N. (2016). Vulnerability Discovery Modeling and Weighted Criteria Based Ranking, Journal of the Indian Society for Probability and Statistics, 1–10.
[10] Bhatt, N., Anand, A., Yadavalli, V.S.S. and Kumar, V. (2017). Modeling and Characterizing Software Vulnerabilities, International Journal of Mathematical, Engineering and Management Sciences (IJMEMS), 2(4), 288–299.
[11] Anand, A., Das, S., Aggrawal, D. and Klochkov, Y. (2017a). Vulnerability discovery modelling for software with multi-versions, Advances in Reliability and System Engineering, Springer, Cham, 255–265.
[12] Beattie, S., Arnold, S., Cowan, C., Wagle, P., Wright, C. and Shostack, A. (2002 November). Timing the Application of Security Patches for Optimal Uptime, LISA, 2, 233–242.
[13] Jiang, Z. and Sarkar, S. (2003 December). Optimal software release time with patching considered, Workshop on Information Technologies and Systems, Seattle, WA, USA.
[14] Arora, A., Caulkins, J.P. and Telang, R. (2006). Research note – Sell first, fix later: Impact of patching on software quality, Management Science, 52(3), 465–471.
[15] Das, S., Anand, A., Singh, O. and Singh, J. (2015). Influence of Patching on Optimal Planning for Software Release and Testing Time, Communications in Dependability and Quality Management, – An International Journal, Serbia, 18(4), 81–92.
[16] Deepika, Anand, A., and Singh, N. (2016). Software Reliability Modeling Based on in-house and Field Testing, Communications in Dependability and Quality Management- An International Journal, Serbia, 19(1), 74–84.
[17] Anand, A., Agarwal, M., Tamura, Y. and Yamada, S. (2017b). Economic impact of software patching and optimal release scheduling, Quality and Reliability Engineering International, 33(1), 149–157.
[18] Anand, A., Gupta, P., Klochkov, Y. and Yadavalli, V.S.S. (2018). Modeling Software Fault Removal and Vulnerability Detection and Related Patch Release Policy, System Reliability Management: Solutions and Technologies, 19–34.
[19] Kaur, J., Anand, A. and Singh, O. (2019). Modeling Software Vulnerability Correction/Fixation Process Incorporating Time Lag, Recent Advancements in Software Reliability Assurance, 39–58.
[20] Ohtera, H. and Yamada, S. (1990). Optimal allocation and control problems for software-testing resources, IEEE Transactions on Reliability, 39(2), 171–176.
[21] Yamada, S., Ichimori, T. and Nishiwaki, M. (1995). Optimal allocation policies for testing-resource based on a software reliability growth model, Mathematical and Computer Modelling, 22(10–12), 295–301.
[22] Xie, M. and Yang, B. (2001). Optimal testing-time allocation for modular systems, International Journal of Quality & Reliability Management, 18(8), 854–863.
[23] Huang, C.Y., Lo, J.H., Kuo, S.Y. and Lyu, M.R. (2004). Optimal allocation of testing-resource considering cost, reliability, and testing-effort, In proceedings of 10th IEEE Pacific Rim International Symposium on Dependable Computing, IEEE: 103–112.
[24] Jha, P.C., Gupta, D., Anand, S. and Kapur, P.K. (2006). An imperfect debugging software reliability growth model using lag function with testing coverage and related allocation of testing effort problem, Communications in Dependability and Quality Management- An International Journal, Serbia, 9(4), 148–165.
[25] Kapur, P.K., Chanda, U. and Kumar, V. (2010). Dynamic allocation of testing effort when testing and debugging are done concurrently, Communications in Dependability and Quality Management- An International Journal, Serbia, 13(3), 14–28.

[26] Kumar, V. and Sahni, R. (2016a). An effort allocation model considering different budgetary constraint on fault detection process and fault correction process, Decision Science Letters, 5(1), 143–156.
[27] Kumar, V., Mathur, P., Sahni, R. and Anand, M. (2016b). Two-dimensional multi-release software reliability modeling for fault detection and fault correction processes, International Journal of Reliability, Quality and Safety Engineering, 23(3), 1640002.
[28] Cao, Z., Lin, J., Wan, C., Song, Y., Zhang, Y. and Wang, X. (2017). Optimal cloud computing resource allocation for demand side management in smart grid, IEEE Transactions on Smart Grid, 8(4), 1943–1955.
[29] Kapur, P.K., Jha, P.C. and Bardhan, A.K. (2003). Dynamic programming approach to testing resource allocation problem for modular software, Ratio Mathematica, 14(1), 27–40.
[30] Anand, A., Das, S., Singh, O. and Kumar, V. (2019, February). Resource Allocation Problem for Multi Versions of Software System, In 2019 Amity International Conference on Artificial Intelligence (AICAI), IEEE: 571–576.
[31] Bhatt, N., Anand, A. and Aggrawal, D. (2019). Improving system reliability by optimal allocation of resources for discovering software vulnerabilities, International Journal of Quality & Reliability Management.

Shinji Inoue and Shigeru Yamada

# 2 Debugging process modeling for quality/reliability assessment of software system

**Abstract:** Debugging activities in a testing phase of software development are important for eliminating software faults remained and for shipping highly reliable software system to the user. Therefore, reflecting the debugging situation in testing activities on software reliability assessment must be one of the useful approaches for managing the software development project with accurate information on software quality/reliability and conducting quality-oriented software management. We introduce two types of software fault debugging-oriented stochastic modeling approaches for conducting mathematical model-based quality/reliability assessment. These modeling approaches discussed in this chapter are expected to improve the accuracy of model-based software reliability assessment.

**Keywords:** software reliability, reliability growth models, debugging process, infinite server queueing, phase-type distribution

## 2.1 Introduction

Software fault debugging in a testing phase is regarded as the final activity for eliminating software faults still remained in the software system and for shipping highly reliable software systems. Generally, a software fault is debugged through the software failure observation, cause analysis, and software fault removal/elimination activities. The growth of the reliability can be observed during the testing activities as the results of the perfect software fault removal activities. That is, the software debugging activities influence on the growth process observed.

As the widely applied assessment technology, software reliability growth models [1, 2] are mathematical models for describing the growth process observed, assessing the software reliability quantitatively, and predicting the future trend of the software reliability growth process. Furthermore, these models provide us with some useful information on quality-oriented software development management, such as optimal shipping time estimation and statistical testing process control. However, most of the models seem like black box type models, which do not incorporate the effect of the debugging process mentioned earlier in the testing activities.

**Shinji Inoue,** Kansai University, Osaka, Japan
**Shigeru Yamada,** Tottori University, Tottorri, Japan

https://doi.org/10.1515/9783110619058-002

In our knowledge, Yamada and Osaki [3] have been first proposed a nonhomogeneous Poisson process-based stochastic model [4] describing the so-called S-shaped reliability growth process considering the two-stage software fault debugging processes, such as the software failure occurrence and the fault isolation processes. Concretely, this model has been developed by focusing on the behavior of the expected counts for software failures occurred and for software faults perfectly debugged by using the differential equations for the mean value function of the nonhomogeneous Poisson process. However, it is difficult to see the uncertainty of software failure occurrence and the fault isolation in the debugging process directly. And there is the lack of flexibility in modeling, that is, it is difficult to expand this model if we need to add other activities of the debugging process and if we need to tailor this model for adjusting to the testing environment managed.

We introduce two types of debugging process-oriented modeling approaches – infinite server queueing and phase-type modeling approaches – for software reliability assessment with debugging activities. These modeling technologies help us to develop a plausible model adjusting to the testing environment managed and to clearly understand the structure and the uncertainties in the debugging process.

## 2.2 Infinite server queueing modeling approach

This modeling approach is for describing the uncertainty in software fault debugging processes, which consists of software failure occurrence and fault-removal processes. This modeling approach is regarded as an extended model of Yamada delayed S-shaped model [3] because this modeling approach describes the uncertainties in the software failure occurrence phenomenon and in fault-removal time flexibly based on the basic assumption of their model, in which the debugging activities consists of the two-stage debugging processes. For the infinite server queueing approach, we assume the following situations being related to the debugging activities:

- The total number of software failure detected follow a nonhomogeneous Poisson process with mean $H(t)$.
- A software fault is completely debugged through the two-stage processes. That is, the software failure occurred is detected through the observation process, then the fault caused this failure is debugged through the removal processes. The fault removal process contains the failure analysis activities.
- The fault removal times are randomly behaved independently and identically with the cumulative probability distribution function $G(t)$.

Following the basic assumptions mentioned earlier, we formulate the system for describing software reliability growth process. Figure 2.1 shows the system configuration for this infinite server queueing system based on the above-mentioned assumptions.

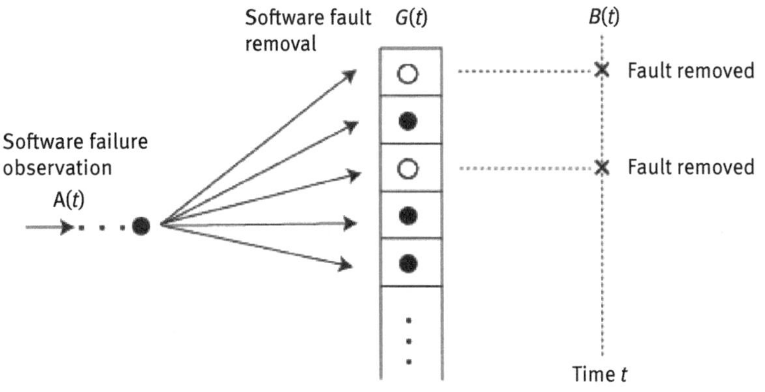


**Figure 2.1:** Configuration of infinite server queueing system.

Now we introduce the stochastic or counting process $\{A(t),\ t \geq 0\}$, which is the cumulative number of software failures observed during (0, *t*] and we also introduce another process $\{B(t),\ t \geq 0\}$ denoting the number of faults removed over (0, *t*]. Considering that the testing activities start at $t = 0$, the stochastic process $B(t)$ is formulated as

$$\Pr\{B(t) = b\} = \sum_{i=0}^{\infty} \Pr\left\{\frac{B(t) = b,\ A(t) = i}{A(t) = i}\right\} \frac{H(t)^i}{i!} \exp[-H(t)]. \tag{2.1}$$

The conditional probability in eq. (2.1), which is the probability that *b* software faults are perfectly/completely debugged during (0, *t*] given that *i* software failures have been observed over (0, *t*], is given by assuming $c(t)$ representing the probability that a software fault causing an observed software failure is removed by time *t*. That is, the conditional probability is given as

$$Pr\left\{\frac{B(t) = b,\ A(t) = i}{A(t) = i}\right\} = \binom{i}{b} c(t)^b \{1 - c(t)\}^{i-b}, \tag{2.2}$$

where $c(t)$ is given by

$$c(t) = \int_0^t \frac{G(t-y)}{H(t)} dH(y). \tag{2.3}$$

In eq. (2.3), $c(t)$ is formulated by following the Stieltjes convolution of the software fault removal time distribution and the conditional distribution of the time of a software failure given $A(t) = i$. It is worth mentioning that this conditional software failure occurrence time distribution per software failure is derived by following the notion of conditional arrival times distribution in infinite server queueing theory [4]. Substituting eqs. (2.2) and (2.3) into (2.1), we can rewrite eq. (2.1) as

$$\Pr\{B(t)=b\} = \frac{1}{b!}\left[\int_0^t G(t-y)dH(y)\right]^b \exp\left[-\int_0^t G(t-y)dH(y)\right]. \tag{2.4}$$

In eq. (2.4)), we can see the process $\{B(t),\ t \geq 0\}$ can be essentially treated as the nonhomogeneous Poisson process, and its expectation is finally formulated by the appropriate functions of $G(t)$ and $H(t)$, respectively.

As shown in eq. (2.4), several functions can be developed for the expectation of the stochastic process $\{B(t),\ t \geq 0\}$, which follows the nonhomogeneous Poisson process, by assuming the suitable functions for the $G(t)$ and $H(t)$, respectively. That is, this modeling approach enables us to explain the details on the physical interpretation on the software debugging process. For example, if we assume that the behavior of the fault removing time follows $G(t) = 1 - \exp[-\theta t])$, which is the exponential distribution with parameter $\theta$ $(>0)$, and the software failure observations follow a nonhomogeneous Poisson process with mean $H(t) = \omega(1 - \exp[-\theta t])$, which is the well-known exponential (Goel–Okumoto) model [1, 2], we can obtain

$$\int_0^t G(t-y)dH(y) = \omega[1 - (1+\theta t)\exp[-\theta t]], \tag{2.5}$$

which is the same mathematical structure as Yamada-delayed S-shaped model. In eq. (2.5), $\omega$ is the expected initial content in the software system and $\theta$ is the software failure occurrence and software fault-removal rates. As shown in eq. (2.5), we can develop several types of models by reflecting actual situations on software fault debugging activities.

## 2.3 Phase-type modeling approach

A phase-type modeling approach is regarded as one of the modeling techniques for generalizing software reliability growth models proposed so far and for developing a more plausible model, which describes a well observed software reliability growth data [5]. This approach can be applied for describing software debugging process more flexibly than the infinite server queueing approach because possible software debugging processes can be described flexibly by using a continuous-time Markov chain. It should be noted that the infinite sever queueing approach describes only the software debugging process developed by the two debugging processes.

For overviewing this approach, the following basic modeling assumptions for the phase-type modeling approach has been introduced:

- The software contains $\Omega_0$ software faults before testing, and $\Omega_0$ is a random variable and takes a nonnegative integer.
- The software failure observation and fault-removing processes are contained as the successive debugging process. The completion time for the successive process follows an independent and identical cumulative probability distribution function $E_{PH}(t)$.
- Any faults are not introduced within the process and faults causing observed software failure are perfectly/completely debugged through the debugging process.

Now we define the process $\{B_{PH}(t),\ t \geq 0\}$ denoting the number of faults removed over (0, *t*], and we see that the process follows

$$Pr\{B_{PH}(t) = b\} = \sum \binom{i}{b} E_{PH}(t)^b \{1 - E_{PH}(t)\}^{i-b} \Pr\{\ \Omega_0 = i\}, \tag{2.6}$$

from the assumptions mentioned earlier. Assuming $\Omega_0$ obeys a Poisson distribution with mean $\alpha\ (>0)$, the process $\{B_{PH}(t),\ t \geq 0\}$ in eq. (2.6) can be rewritten as

$$Pr\{B_{PH}(t) = b\} = \frac{1}{b!}\{\alpha E_{PH}(t)\}^b \exp[-\alpha E_{PH}(t)] \quad (b = 0,\ 1,\ 2,\ \ldots). \tag{2.7}$$

From eq. (2.7), we see that the stochastic process $\{B_{PH}(t),\ t \geq 0\}$ has essentially the same mathematical structure as eq. (2.4). However, the process has a different time-dependent expectation, that is, $\alpha E_{PH}(t)$.

Let us consider $E_{PH}(t)$ expressing the uncertainty of the completion time of the successive processes, which contains the software failure observation and perfect fault removing. Regarding the debugging process being related to the fault removal after software failure observation, it is possible to consider several processes developed by not only one process but also by multiple or parallel processes. For describing such debugging process flexibly, we apply a phase-type probability distribution [6] for $E_{PH}(t)$. The phase-type distribution describes the time uncertainty from the initial state to the absorption in the continuous-time absorbing Markov chain. Figure 2.2 shows a basic configuration of the continuous-time absorbing Markov chain. It is worth mentioning that the phase means the transient states in the absorbing Markov chain.

Now we mention the mathematical discussion about the absorbing continuous-time Markov chain, which consists of $V_T = \{1,\ 2,\ \ldots,\ n\}$, the transient states set, and the absorbing state $V_A = \{n+1\}$. Generally, the instantaneous behavior within the states can be expressed by an infinitesimal generator $\boldsymbol{I}$ as

$$I = \begin{pmatrix} S & A_1 \\ 0 & 0 \end{pmatrix}, \tag{2.8}$$

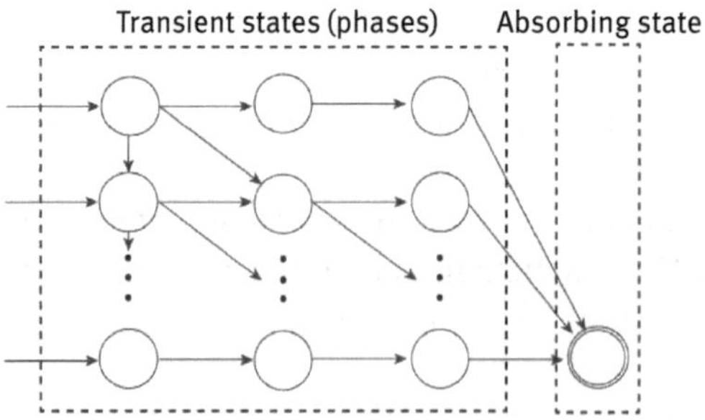


**Figure 2.2:** Configuration of continuous-time absorbing Markov chain.

where $S$ is $n \times n$ submatrix representing the instantaneous transition rate within the set of $V_T$ and $A_1$ is the column vector representing the rate from $V_T$ to $V_A$, respectively. Further, 0 in eq. (2.8) is the row vector consisting of 0 elements representing no transition from $V_A$ to $V_T$ and no transition within the absorbing state. As mentioned earlier, the phase-type distribution expresses the randomness of the time from the initial state to the absorption. Therefore, we can consequently obtain the cumulative probability distribution for the time to absorption from the initial states as

$$E_{PH}(t) = \Pr\{T \le t\} = 1 - \pi_0 \exp[St]1 \quad (t \ge 0), \tag{2.9}$$

where $\pi_0$ represents the initial state vector and 1 is the column vector whose elements is 1. From the modeling framework in eqs. (2.7)–(2.9), we can obtain stochastic models for software reliability assessment by developing suitable continuous-time absorbing Markov chain reflecting software fault debugging processes.

Of course, this modeling approach gives us analysis method for the uncertainty of the software fault debugging processes of several models proposed so far. For example, the debugging process of the delayed S-shaped model can be analyzed by following this modeling approach. Figure 2.3 shows the diagram expressing state transitions of this model. The diagram shown in Figure 2.3 consists of the two processes: (1) the software failure observation process and (2) the fault-removal or fault isolation process, respectively. From Figure 2.3, the infinitesimal generator for the absorbing Markov chain is

$$I = \begin{pmatrix} -d & d & 0 \\ 0 & -d & d \\ 0 & 0 & 0 \end{pmatrix}, \tag{2.10}$$

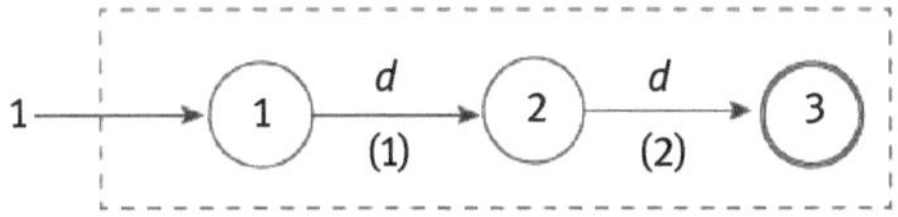


**Figure 2.3:** Absorbing Markov chain for the delayed S-shaped model.

and $\pi_0 = (1 \quad 0 \quad 0)$. Further, we can see that $V_T = \{1,\ 2\}$, $V_A = \{3\}$ and

$$S = \begin{pmatrix} -d & d \\ 0 & -d \end{pmatrix}, \tag{2.11}$$

respectively. Then, we can obtain the phase-type probability distribution as

$$E_{PH}(\mathrm{t}) = 1 - (1 \quad 0 \quad 0)\exp\left[\begin{pmatrix} -d & d \\ 0 & -d \end{pmatrix} t\right]\begin{pmatrix} 1 \\ 1 \\ 1 \end{pmatrix} = 1 - (1 + dt)\exp[-dt]. \tag{2.12}$$

Substituting eq. (2.12) into eq. (2.7), the mean value function reflecting the software faults debugging processes can be obtained as $\alpha E_{PH}(t) = \alpha\{1 - (1 + dt)\exp[-dt]\}$, which is essentially the same mathematical structure as the delayed S-shaped model. Several types of model reflecting debugging processes can be developed by following the procedure mentioned earlier. However, it is observed that the number of parameters might increase if the details in the debugging process and different transition rates within the chain are considered. Therefore, useful procedures for estimating the parameters, such as the EM algorithm needs to be investigated [5].

## 2.4 Conclusion

We slightly introduce the software fault debugging process-oriented reliability growth modeling approaches, such as infinite server queueing and phase-type modeling approaches. These two modeling approaches yield software reliability growth modeling frameworks reflecting the uncertainties of software fault debugging processes on software reliability assessment and prediction. Some other models could be developed based on these modeling frameworks with the difficulty of fault debugging or some attributes being related to the debugging activities. Further, these debugging process-oriented modeling techniques lead to quality-based software development project management by analyzing the efficiency of debugging activities in the testing activities, which are cost- and time-consuming activities.

# Bibliography

[1] Pham, H. (2000). Software Reliability, Springer-Verlag, Singapore.
[2] Yamada, S. (2014). Software Reliability Modeling – Fundamentals and Applications, Springer Japan, Tokyo, Japan.
[3] Yamada, S. and Osaki, S. (1985). Software reliability growth modeling: Models and applications, IEEE Transactions on Software Engineering, SE-11, 1431–1437.
[4] Ross, S.M. (2010). Introduction to Probability Models, Elsevier, San Diego, CA, USA.
[5] Okamura, H. and Dohi, T. (2016). Phase type software reliability model: parameter estimation algorithms with grouped data, Annals of Operations Research, 244, 177–208.
[6] Buchholz, P., Kriege, J. and Felko, I. (2014). Input Modeling with Phase-Type Distributions and Markov Models: Theory and Applications, Springer, Cham, Switzerland.

Shivani Gupta and Vandana Gupta

# 3 Availability analysis of vehicular cloud computing

**Abstract:** Vehicular cloud computing is a new paradigm that makes use of cloud computing resources to overcome the restraints of vehicular computing. It allows the sharing of resources such as storage capacity, computational power, and Internet connectivity from those vehicles in which these resources remain idle for long hours (in parking or heavy traffic jams). This chapter presents an availability study of the vehicular cloud. Due to its multilayered architecture, composite modeling technique is desired for availability analysis of vehicular clouds. Distinct models are developed for each subsystem using reliability block diagrams and semi-Markov process, and the models are then combined to evaluate the availability of the complete system. Two different sensitivity analysis techniques (partial derivatives and percentage difference) to determine the parameters that cause the greatest impact on the availability of the vehicular cloud are also applied. The analysis reflects that the availability can be improved by aiming on reduced set of parameters that affect the availability most.

**Keywords:** Vehicular cloud computing, availability analysis, composite modeling, reliability block diagram, semi-Markov process, sensitivity analysis

## 3.1 Introduction

Vehicular ad hoc network (VANET) communicates information (in the form of data packets) regarding incidents such as congestions on road, accidents, and other road conditions that are away from the driver's knowledge. This significantly enhances the reliability and effectiveness of the transportation system. The amalgamation of numerous devices such as global positioning system (GPS), general packet radio service (GPRS), various sensors, and interfaces has contributed in making the vehicles smart. Apart from providing the above-mentioned security services, smart vehicles also facilitate the users with various infotainment services (e-mail, video downloading and sharing, web browsing, etc.). Due to the increase in number of applications, the demand for resources such as storage capacity, internet connectivity, and computational power is increasing rapidly.

**Gupta Shivani, Vandana Gupta,** University of Delhi, Delhi, India

https://doi.org/10.1515/9783110619058-003

In such situations, a single vehicle is sometimes unable to provide all the required resources. At the same time, the computational resources of some vehicles remain idle for long hours (in parking or heavy traffic jams). Therefore, sharing of these resources is suggested to utilize the idle resources, and also to provide satisfactory services to the users. This sharing of under-utilized resources is termed as vehicular cloud computing (VCC). It is inspired by cloud computing (CC) and mobile cloud computing (MCC). CC is the sharing of computing services such as storage, networking, servers, databases, intelligence, and analytics through the internet. MCC can be simply characterized as an infrastructure in which both the data processing and data storage take place outside the mobile device. It is based on the perception that businesses can operate by borrowing the necessary software and infrastructure without spending money in purchasing them.

In a VCC network, vehicles connect with each other to either share their resources or to rent out. The roadside infrastructure is also a part of VCC as they can also rent out their resources. Each vehicle in a VCC network can act as both a service user and as a service provider. As an example, a VCC network can be created by using computing resources of the vehicles parked in the parking of a company. In this VCC network, the company doesn't need to spend money in buying computing resources, and the owners also get some money for sharing their idle resources. In this way it helps both the company as well as the owners of the vehicles. Similarly, a VCC network can also be formed in traffic jams to update people stuck in traffic about traffic conditions and also to transmit the data efficiently.

These days every automobile-making company has completely accepted that cloud is necessary to provide competitively distinctive services and features for current and future users. As per Gartner, quarter billion vehicles will be connected on road by 2020 [1]. Due to such huge numbers, it is necessary to deliver trustworthy services, that is, dependable services.

The term dependability is used to define the system's ability to facilitate the users with those services that can be trusted justifiably within a time period [2]. Dependability is critical for both service providers and users. It is a comprehensive theory that includes measures of system's availability, reliability, maintainability, security, integrity, and so on. [2–4]. Availability is one of the key attributes of dependability that directly relates to the fraction of time the system is in functioning state [5]. There are numerous types of models that are employed to evaluate the analytical measures, which are analogous to system's availability. These models are broadly divided into two categories: state-space models and nonstate-space models. Markov chains and stochastic petri nets are the examples of state-space models whereas, fault trees, reliability block diagrams (RBDs) are the examples of nonstate-space models. State-space models facilitate the portrayal of intricate connections among system's components, while nonstate-space models acknowledge a much compact representation of the system. Hierarchical combination of both the models, that is, state-space models and nonstate-space models provide us with the best of both the modeling approaches [6].

In this chapter, we adopt the hierarchical modeling approach to perform the availability analysis of VCC. Distinct models are proposed for each subsystem of the VCC, and the developed state-space and nonstate-space models are then combined to evaluate the availability for the complete VCC network. Sensitivity analysis is also performed to determine those parameters that impact the steady-state availability the most, that is, a small change in the value of the input parameter will significantly affect the availability. We have used two different techniques, namely, partial derivative technique and percentage difference technique for the sensitivity analysis. Two different techniques are used so that the results obtained from one technique can be verified by the other.

## 3.2 Literature review

Due to the advancement in the field of communication and computational technologies, automobile industry is experiencing a progressive change. CC is a paradigm that can utilize the advance computing resources available in the smart vehicles. The integration of CC with the vehicular networking develops a new paradigm, known as VCC. Authors in [7–9] have explained the concept of VCC, which facilitates the sharing of internet connectivity, storage, and computing power among the users. Plenty of research is available in literature giving an overview of the VCC's architecture, features, applications, and security challenges [10–13].

In [14] authors have suggested a new scheme to enhance the availability of cloud services and to minimize the end to end latency in vehicular communication. Authors in [15] have proposed a message confirmation system to make the communication more reliable in VCC. In [16] authors have studied the effect of social relation on mobile video offloading services. They have proposed a resource allocation scheme based on continuous-time Markov decision process. It enhances the quality of entertainment demands. Authors in [17] have proposed a resource scheduling and task allocation algorithm to accomplish numerous task requests coming from the VCC users in an adequate manner. It upgrades the quality of service of the VCC network. In [18] authors have highlighted the issue of inefficiency in information transmission in vehicular cloud. They have proposed a network coding method that helps in improving the reliability and efficiency in the information transmission. Authors in [19] have addressed the issue of security and privacy in VCC, and developed a security and privacy-aware information transmission environment between the cloud infrastructure and vehicular nodes. Authors in [20] have dealt with the dynamic nature of the vehicular cloud. They have proposed a vehicular cloud model that possesses different task requests types and different resource capability types. It reduces the failure rate in the vehicular cloud. In all these papers discussed earlier, we feel that availability study of the VCC architecture is immensely ignored. This motivated us to study the availability of a VCC network using analytical modeling approach. None of the authors have focused on the availability analysis of the vehicular cloud. The next section discusses about the major contribution of this chapter.

### 3.2.1 Our contribution

The major contributions of this chapter are listed as follows:

- In this chapter the availability of the VCC architecture is evaluated using hierarchical modeling approach.
- Sensitivity analysis is also performed using two different techniques to compute the effect of each input parameter on the steady-state availability. Two different techniques have been used so that the results obtained from one technique can be verified by the other.
- With the help of sensitivity analysis those parameters that cause great impact on the availability of the VCC architecture are identified.

The chapter is summarized as follows: In Section 3.3 the architecture of VCC is discussed. In Section 3.4 different availability models for all the components of VCC architecture are proposed. Further, in Section 3.5 steady-state availability of the whole VCC architecture is evaluated. Also, sensitivity analysis is performed through two different techniques. Finally, the chapter is concluded in Section 3.6.

## 3.3 Vehicular cloud computing architecture

The VCC architecture is divided into three layers: *Inside Vehicle, Communication layer, and Cloud* [21]. The hierarchical architecture of VCC is shown in Figure 3.1. In the inside vehicle layer, the major component considered is the On-board unit (OBU). The OBU consists of a control unit (CU), GPS, GPRS, various input/output interfaces, and some sensors (such as body sensors, environmental sensors, and driver's behavior recognition). The second layer of this architecture is the communication layer. Two types of communication are possible in vehicular communication network: (i) vehicle-to-vehicle communication (V2V), and (ii) vehicle-to-infrastructure communication (V2I). In V2V communication vehicles communicate with each other through Dedicated short-range communication (DSRC), whereas in V2I communication vehicles communicate with the roadside infrastructure (street light, traffic signal poles, etc.) into which the equipment for wireless network such as 3G or Wi-Fi is installed.

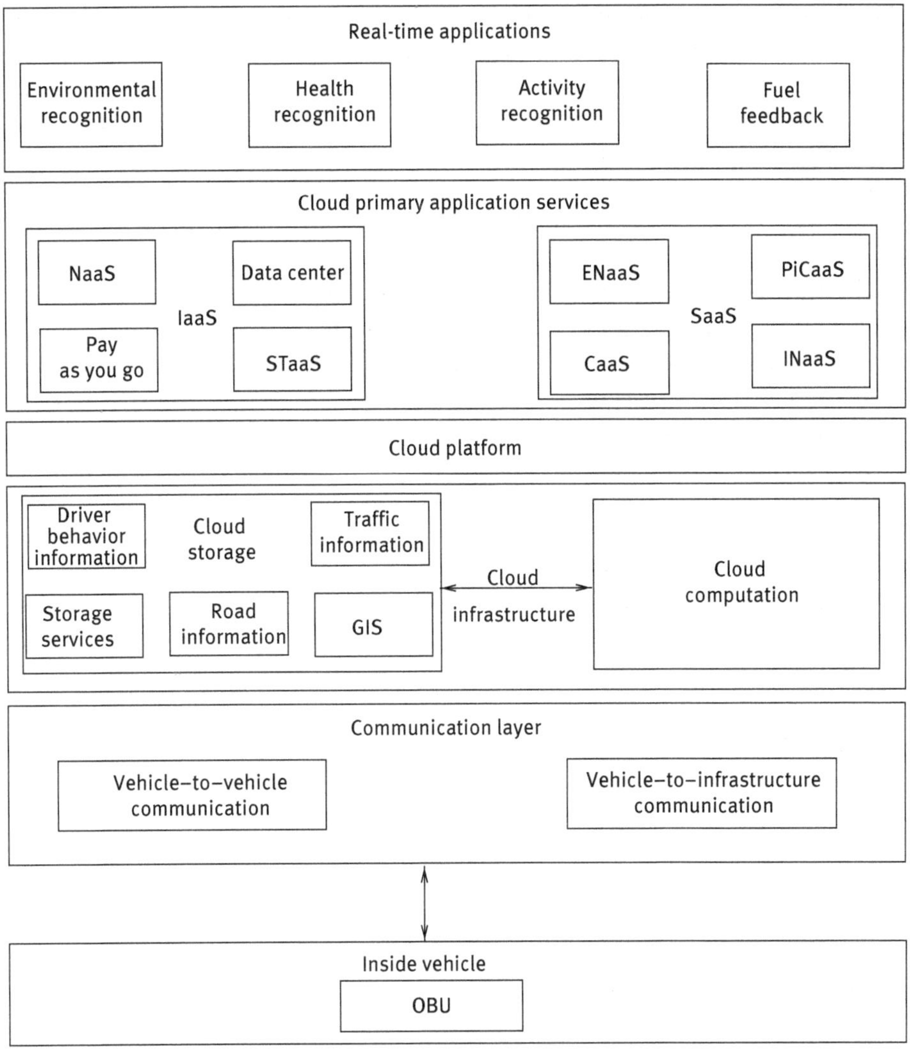


**Figure 3.1:** Vehicular cloud computing architecture.

The upper most layer of this architecture is the *cloud*, which is further divided into three sublayers: *cloud infrastructure*, *cloud platform*, and *cloud primary and real-time application services*. Cloud infrastructure is divided into two parts: *cloud storage* and *cloud computation*. The data collected by the inside vehicle layer is stored in this cloud storage. Numerous government and private firms, especially the police department can use this stored data in the *cloud* for various studies [21]. Cloud computation performs computing tasks using the data stored in cloud storage. It can carry out complex calculations in nominal time. In the cloud primary and real-time application services, numerous applications and services are provided, which can be directly accessed by the users. Health recognition, environmental recognition, and fuel feedback are some examples of real-time applications provided by the cloud. The primary application services in the cloud are broadly divided into two categories: Infrastructure as a Service (IaaS) and Software as a Service (SaaS). IaaS includes those applications in which some virtual as well as physical storage options, servers, or networks are required such as Network as a Service (NaaS), Storage as a Service (STaaS), Data Center, pay-as-you-go, whereas SaaS is related to infotainment and driver comforting applications such as Entertainment as a Service (ENaaS), Cooperation as a Service (CaaS), Pictures on wheels as a Service (PiCaaS), and so on [22]. NaaS can also be delivered as a SaaS.

In this chapter, in *cloud platform layer*, we majorly focus on IaaS model. It is the most fundamental level of cloud services. An open source IaaS service provider is EUCALYPTUS – Elastic Utility Computing Architecture Linking Your Programs to Useful Systems. Its interface is appropriate for the economic services such as Amazon EC2 (Elastic Compute Cloud) and Amazon S3 (Simple Storage Service) [23]. The architecture of EUCALYPTUS consists of five high-level components: cloud controller (CLC), cluster controller (CC), node controller (NC), storage controller (SC), and Walrus. Each of these components has its own web interface. These components are explained as follows [24]:
- **CLC** is the entry point of the complete cloud infrastructure. Besides receiving requests from client side through user interface, it also interacts with the remaining components of the EUCALYPTUS.
- **CC** is the entry point of the *cluster*. It interacts with both NCs and CLC. Its major tasks are to determine which NC will perform the incoming service requests, to collect information regarding the nodes that constitute its cluster, and to manage the virtual network overlay.
- **NC** is possessed by every physical node that is assumed to run virtual machines (VM). It manages the implementation, analysis, and completion of VM instances on the hosts where it runs.
- **SC** facilitates with constant block storage that is used by VM instances.
- **Walrus** provides a mechanism for storing VM images. It is an interface adaptable file-based data storage service. It also allows user to flow data inside and outside the cloud.

In this chapter, we consider the EUCALYPTUS-based cloud environment to develop the availability model and assume that it has three clusters. Each cluster has one CC, one SC, and several NC. Components of each cluster communicate with the CLC and Walrus in order to manage service requests.

## 3.4 Availability models for vehicular cloud

The availability model for the VCC architecture discussed in Section 3.3 is shown in Figure 3.2. The architecture is modeled through an RBD. RBD is a tool to model complex systems. The different blocks of the RBD represent the various components of the VCC such as OBU, V2I, V2V, Cloud Storage, CLC, and CC subsystems. Once the blocks are figured correctly, various system measures such as reliability, availability, MTBF, and failure rate, can be calculated from the RBD. To evaluate the availability of vehicular cloud from the RBD shown in Figure 3.2, distinct availability models for each block is developed, and then the results obtained from the submodels are combined to evaluate the system's availability.

Hence, essentially in this section, the availability models developed for each component of the VCC network is discussed, and then the availability of the complete VCC network by the amalgamation of all the submodels will be evaluated.

### 3.4.1 Availability model for on-board unit

The availability model for the OBU is shown in Figure 3.3. The model is mapped through an RBD. We have considered various components of OBU that can affect its availability. A brief introduction of its components is as follows [25]:

- **CU** is one of the major components of OBU. It facilitates the OBU in maintaining records regarding the incidents disclosed, processed, and transmitted to it by the roadside infrastructure. It also allows the OBU to coordinate and connect among its external devices.
- **GPS** provides the geolocation and time information to the users.
- **GPRS** enables wireless communication in vehicular environment. It also provides Internet facility.
- **Input/output (I/O) interface** provides communication links between internal system and input–output devices.
- **Various sensors** are available inside and outside the vehicles. They collect information regarding pressure and temperature inside the car, driver's behavior, and mood as well as send all these data to cloud storage or to the software for various real-time application services.

We consider a series combination of all the components of OBU because unavailability of any one of the components affects the availability of the OBU. The closed-form equation for availability of the RBD shown in Figure 3.3 is given as

$$A_{\mathrm{OBU}} = A_{\mathrm{CU}} \times A_{\mathrm{GPS}} \times A_{\mathrm{GPRS}} \times A_{\mathrm{I/O}} \times A_{\mathrm{Sensors}}. \tag{3.1}$$

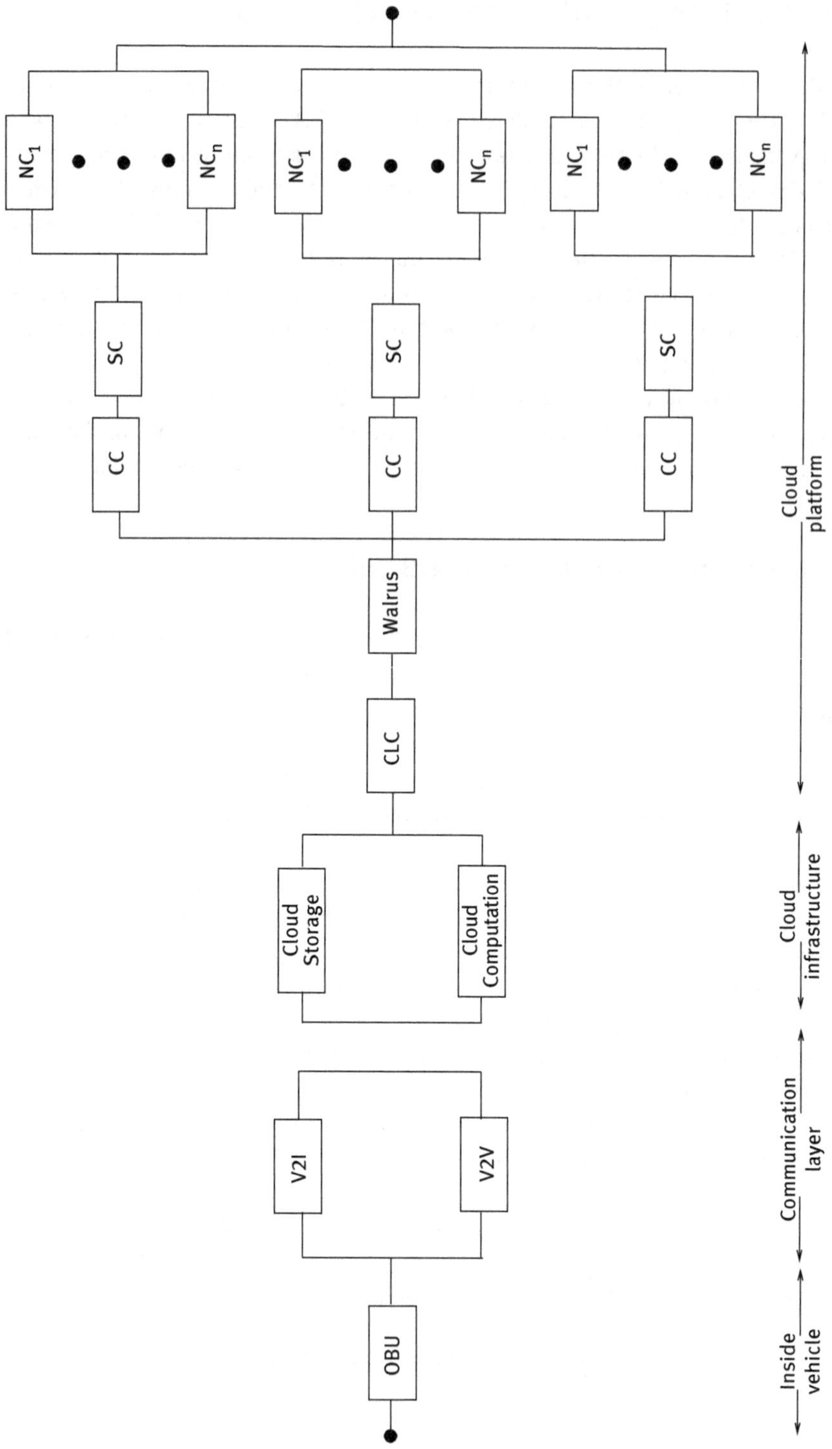


**Figure 3.2:** Availability model for vehicular cloud architecture.

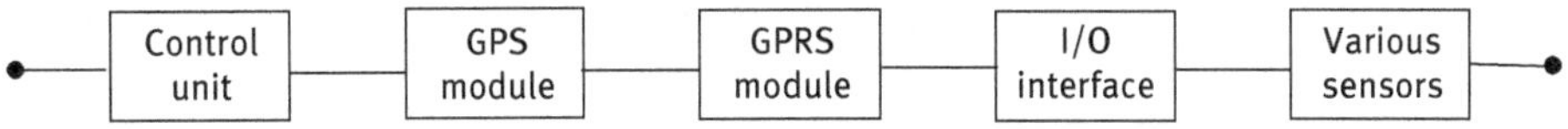


**Figure 3.3:** Availability model for OBU.

Each term $A_i$ in the above-mentioned equation represents the availability of the $i$th component, where $i$ ε {CU, GPS, GPRS, I/O, Sensors}. The availability of the $i$th component is calculated using the formula given as follows:

$$\text{Availability} = \frac{\text{MTBF}}{\text{MTBF} + \text{MTTR}}. \tag{3.2}$$

Note that in this chapter the values of MTBF and MTTR are not evaluated by us and are actually borrowed from the literature.

### 3.4.2 Availability model for vehicle-to-vehicle communication

In V2V communication, vehicles communicate with each other. The availability of V2V communication network depends upon the number of OBUs (i.e., vehicles) in functioning state among all the OBUs that can communicate within the transmission range [25]. Let us consider that the transmission range of an OBU is $R$ and within the transmission range there can be at most $N$ vehicles. Therefore, the V2V communication network will be available if there are at least two OBUs in functioning condition among the $N$ OBUs. Considering that the OBUs in all the vehicles are identical, the availability for V2V communication can be mathematically expressed as

$$A_{\text{V2V}} = \sum_{K=2}^{N} \binom{N}{k} A^k{}_{\text{OBU}} (1 - A_{\text{OBU}})^{N-k}, \tag{3.3}$$

where $A_{\text{OBU}}$ can be obtained from eq. (3.1).

### 3.4.3 Availability model for vehicle-to-infrastructure communication

In V2I communication vehicles communicate with the roadside infrastructure. The equipment for wireless network (such as 3G/4G/5G or Wi-Fi) is installed into these roadside infrastructures. In this chapter, we assume that only 5G network is available for communication. Therefore, for the availability of V2I communication network, at least one OBU and the wireless network should be in functioning condition. Hence, the availability of V2I communication network can be mathematically expressed as

$$A_{V2I} = A_{V2V} \times A_{5G}, \tag{3.4}$$

where $A_{5G}$ denotes the availability of the 5G network. $A_{V2V}$ is evaluated using eq. (3.3), and $A_{5G}$ is given as

$$A_{5G} = \frac{\mu_{5G}}{\mu_{5G} + \lambda_{5G}}. \tag{3.5}$$

Here the parameters $\mu_{5G}$ and $\lambda_{5G}$ are the repair and failure rates, respectively, of the 5G network.

### 3.4.4 Availability model for cloud storage

In cloud storage the data is stored in logical pools. The physical storage connects with different servers (sometimes in different locations), and this physical environment is inherited and governed by a hosting company. These companies are responsible for the applicability and achievability of the data and to keep it running. Further, maintaining the integrity and accessibility of data is also their responsibility. Cloud storage services are accessed by the users through application programming interface (API) or by applications that handle the API such as web-based content management systems, cloud storage gateway, or cloud desktop storage [26]. The availability model that represents the cloud storage is shown in Figure 3.4. The model is developed through an RBD. In the RBD model we have considered n locations in parallel since in cloud storage identical data is stored in different servers at different locations for backup. A closed-form equation for the availability of cloud storage is given as follows (Figure 3.4):

$$A_{Storage} = A_{API} \times A_{storage\_pool} \times (1 - (1 - A_{VC\_server}) \times (1 - A_{PS\_server}))^n \tag{3.6}$$

where $A_{API}$, $A_{storage\_pool}$, $A_{VC\_server}$, and $A_{PS\_server}$ represent the availability of API, logical storage pool, virtual compute server, and physical storage server, respectively. The availability of all these components can be evaluated from the formula given in eq. (3.2). The equation is obtained through the general series–parallel equations for RBDs as discussed in [27].

### 3.4.5 Availability model for cloud controller

The CLC is a single component unlike the other components of VCC network discussed earlier. We assume that a redundant CLC is also available in warm stand-by to provide uninterrupted service. Therefore, to evaluate the availability of CLC, we make use of state-space model. The availability model for CLC is represented in Figure 3.5.

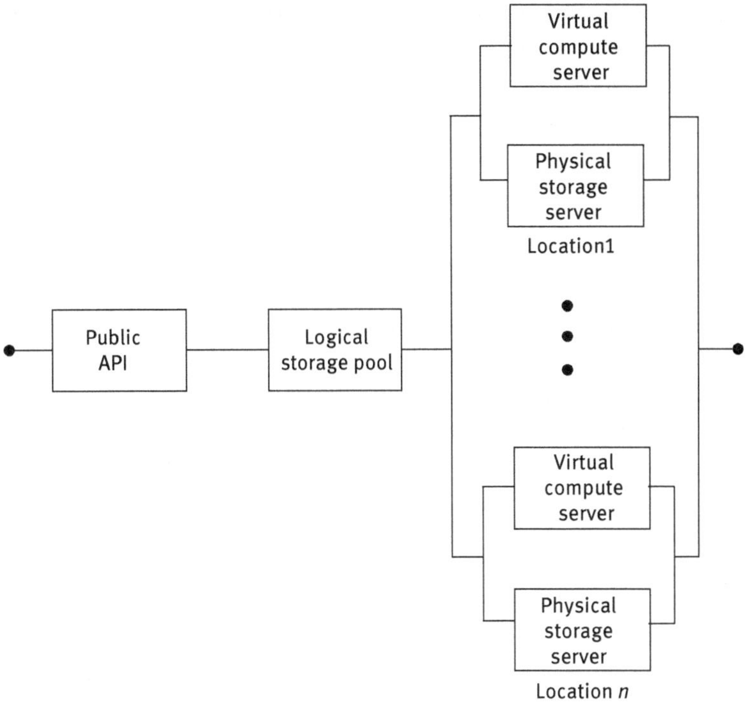


**Figure 3.4:** Availability model for cloud storage.

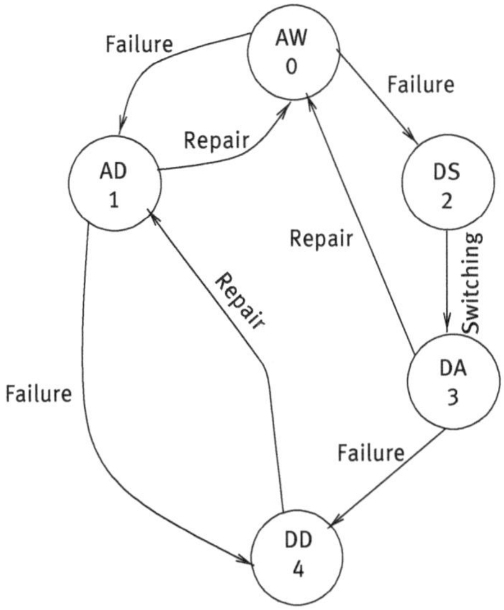


**Figure 3.5:** Availability model for cloud controller and cluster controller.

The proposed state-space model has five states: **AW**, **AD**, **DS**, **DA**, and **DD**. The first letter of the state represents the status of primary CLC and the second letter represents the status of the secondary CLC. The letter **A** denotes active, **W** denotes waiting, **S** denotes switching, and **D** denotes down. A brief description of all the states is discussed as follows:

- State **AW**: Primary CLC is active and secondary CLC is waiting in warm stand-by mode.
- State **AD**: Primary CLC is active and secondary CLC got down while working in warm stand-by mode.
- State **DS**: Primary CLC is down and switching of secondary CLC over primary is in process.
- State **DA**: Primary CLC is down and secondary CLC is active.
- State **DD**: Both the primary and secondary CLCs are in down state.

Initially, the system is in state AW, that is, the primary CLC is active and secondary CLC is in warm stand-by mode. When the primary CLC fails, the system moves to the state DS. Here, a switching of stand-by CLC to primary CLC is performed. Next, the system moves to the state DA, where primary CLC is down and secondary CLC is active. If the secondary CLC got down before the repair of the primary CLC, then the system moves to the state DD. In this model we assume that there is a single repair facility, and the primary CLC repair has preference over the secondary CLC repair. Hence, whenever the system reaches to state DD, there is only one possibility of going to state AD. Also, if the secondary CLC got down in stand-by mode then the system reaches to state AD from state AW.

It is noticeable that the transition from one state to another is affected by multiple factors such as system parameters and it shows random behavior. This random behavior can be modeled by stochastic process such as Markov chain, Poisson process, SMP, and so on. Since the switching time from the primary CLC to the secondary CLC is deterministic, therefore, the time spent in state DS, that is, its sojourn time is nonexponential. Hence, this model cannot be mapped through a simple Markov chain. Further, if we consider the sojourn times of all the states to be exponentially distributed, then this may produce some critical errors. Therefore, in order to capture real-time situations, we consider that the sojourn time of different states may or may not be exponentially distributed, and hence an SMP can be used to model the availability of the CLC. To further facilitate the modeling using an SMP we assume that the Markov property is held at each transition epoch. Hence, we model the state transition diagram for CLC as an SMP [4]. The list of cumulative distributive functions (CDFs) of time spent in each state along with the distribution function and parameters is mentioned in Table 3.1. In the table, $F_{lm}$ denotes the CDF related to the arc $l \rightarrow m$ (where $l, m = 0, 1, 2, 3, 4$).

**Table 3.1:** List of distributions.

| CDF | Distribution | Parameter |
|---|---|---|
| $F_{01}$ | Exponential | $\lambda_{01}$ |
| $F_{02}$ | Exponential | $\lambda_{02}$ |
| $F_{10}$ | Exponential | $\lambda_{10}$ |
| $F_{14}$ | Exponential | $\lambda_{14}$ |
| $F_{23}$ | Deterministic | $T$ |
| $F_{30}$ | Exponential | $\lambda_{30}$ |
| $F_{34}$ | Exponential | $\lambda_{34}$ |
| $F_{41}$ | Exponential | $\lambda_{41}$ |

In the SMP discussed earlier, system is available to the users in all the states except the down state. Therefore, the steady-state availability is obtained as

$$A_{\mathrm{CLC}} = \pi_1 + \pi_2 + \pi_3 = 1 - \pi_4 \tag{3.7}$$

where $\pi_{\mathrm{I}}$ represents the steady-state probability of state $i$ (where $i = 0,1,2,3,4$). These steady-state probabilities are evaluated from the steady-state analysis of the SMP [4; 28].

### 3.4.6 Availability model for cluster controller

Similar to CLC, the CC is also a single component for which warm stand-by replication strategy is assumed. Therefore, the availability model for CC is identical to the availability model for CLC is shown in Figure 3.5. The meaning of all the states of the availability model for CC shown in Figure 3.5 is same as defined for CLC. The sojourn time of all the states also follow the same distribution as in the case of CLC. Hence, the steady-state analysis is also identical for both CLC and CC. Therefore, the steady-state availability of CC is also obtained as

$$A_{\mathrm{CLC}} = \pi_1 + \pi_2 + \pi_3 = 1 - \pi_4 \tag{3.8}$$

where $\pi_{\mathrm{I}}$ represents the steady-state probability of state $i$.

### 3.4.7 Availability of vehicular cloud computing architecture

In this section, we evaluate the availability of the whole VCC architecture by combining the availabilities of the submodels obtained in the earlier sections. A closed-form equation for availability of the VCC system, obtained from the top-level RBD shown in Figure 3.2 is as follows:

$$A_{\text{sys}} = A_{\text{OBU}} \times (1 - (1 - A_{\text{V2I}}) \times (1 - A_{\text{V2V}})) \times (1 - (1 - A_{\text{Storage}}) \times (1 - A_{\text{Comp}}))$$
$$\times A_{\text{CLC}} \times A_{\text{Walrus}} \times (1 - (1 - A_{\text{CC}} \times A_{\text{SC}} \times (1 - \Pi_{i=1}^{n}(1 - A_{\text{NC}_i})^{n}))^{3})$$

Each term $A_x$ in the equation is obtained by solving their corresponding submodels, where $x$ ε {OBU, V2I, V2V, Storage, Comp, CLC, Walrus, CC, SC, $NC_i$}; $i = \overline{1, n}$. The availabilities of the blocks illustrating cloud computation, Walrus, SC, $NC_i$ are calculated by using eq. (3.2).

## 3.5 Numerical illustration and sensitivity analysis

In this section, we perform numerical experiment and evaluate the steady-state availabilities of all the components of the VCC that have been discussed in the earlier section. The values of the input parameters for the OBU are given in Table 3.2. The values of MTBF and MTTR of all the components of OBU are taken from [25]. Further in Table 3.3 values of the input parameters for the communication layer are mentioned. The parameter values for 5G communication are referred from [5]. In Tables 3.4 and 3.5, the values of MTBF and MTTR of all the components of cloud infrastructure layer and cloud platform layer are given. These values are adopted from [24]. Table 3.6 contains the values of the transition rates of the SMP model for CC and CLC as shown in Figure 3.5. Same set of values are considered for both the CC and the CLC model. These values are also taken from [24].

**Table 3.2:** Input parameters for the OBU.

| Component | MTBF (h) | MTTR (h) |
|---|---|---|
| Control unit | 880 | 0.25 |
| GPS | 870 | 0.1667 |
| GPRS | 860 | 0.25 |
| I/O interface | 850 | 2 |
| Various sensors | 850 | 0.25 |

**Table 3.3:** Input parameters for communication layer.

| Parameter | Value |
|---|---|
| $\lambda_{5G}$ | 0.00001 |
| $\mu_{5G}$ | 0.083 |
| $N$ | 50 |

**Table 3.4:** Input parameters for the cloud infrastructure layer.

| Component | MTBF (h) | MTTR (h) |
|---|---|---|
| API | 850 | 2 |
| Storage_pool | 250.1 | 1.12 |
| VC_server | 5,600 | 19.6 |
| PS_server | 3,360 | 38.5 |
| Cloud computation | 789.4 | 0.5 |

**Table 3.5:** Input parameters for the cloud platform layer.

| Component | MTBF (h) | MTTR (h) |
|---|---|---|
| Walrus | 789.4 | 1 |
| SC | 789.4 | 1 |
| NC | 789.4 | 1 |

**Table 3.6:** Input parameters for the CC and CLC.

| Parameter | Value (h) | Parameter | Value (h) |
|---|---|---|---|
| $\lambda_{01}$ | 0.0025 | $\lambda_{30}$ | 1.075 |
| $\lambda_{02}$ | 0.003 | $\lambda_{34}$ | 0.003 |
| $\lambda_{10}$ | 1.075 | $\lambda_{41}$ | 1.075 |
| $\lambda_{14}$ | 0.003 | $T$ | 0.005 |

## 3.5.1 Availability analysis

The availability measures are computed for each submodel discussed in Section 3.4 using the input parameters mentioned earlier. Then the availability of the whole VCC network is evaluated using eq. (3.9). Along with the steady-state availability, the following two measures are also evaluated:

- **Number of nines:** It provides a logarithmic perspective of availability. It is calculated by using the following formula:

$$\text{Number of nines} = -\log_{10}x$$

where $x$ represents the unavailability of the system. If a system is available to the users throughout the year then it is said that it has 5 nines availability or class 5 availability.

- **Downtime:** Downtime provides the total hours in a year for which the system is in down state.

The results of the analysis are summarized in Table 3.7. As per the analysis, the steady-state availability of the VCC network is 0.991501. It has approximately 2 nines availability. The downtime value of 74.5 h/year indicates that the system is unavailable to the users for more than three days in a year. All these results reflect that there is a need to increase the availability of the system.

**Table 3.7:** Availability results.

| Availability | Number of nines | Downtime (h/year) |
|---|---|---|
| 0.991501 | 2 | 74.5 |

In the next section, we have applied different sensitivity analysis techniques to recognize those parameters that can affect the availability of the system.

### 3.5.2 Sensitivity analysis

Sensitivity analysis is a technique that helps in identifying those input parameters that impact the output measure most, that is, a small change in the value of input parameter will significantly affect the output measure. The main objective of sensitivity analysis is to determine those parameters that are the bottlenecks for the steady-state availability. It also helps us to select those parameters that have minimal effect on the steady-state availability, so that we can fairly ignore them. There are numerous methods for performing sensitivity analysis. The fundamental method among all is to change one parameter at a time while keeping the others constant. A sensitivity ranking is obtained after performing sensitivity analysis by noting the changes to the output measure. In this chapter two techniques, namely, partial derivative technique and percentage difference technique are used to perform sensitivity analysis. We are using two different techniques so that the results obtained from one technique can be verified by the other.

### 3.5.2.1 Partial derivative technique

Differential analysis is the basis of many sensitivity analysis techniques [5]. In this technique we simply evaluate the partial derivative of the measure with respect to each input parameter. The sensitivity coefficient SS of the measure Z with respect to the input parameter Θ is evaluated with the help of eqs. (3.10) and (3.11) [29]:

$$S_\theta(Z) = \frac{\partial Z}{\partial \theta} \tag{3.10}$$

$$SS_\theta(Z) = \frac{\partial Z}{\partial \theta}\left(\frac{\theta}{Z}\right) \tag{3.11}$$

Here, in eq. (3.11) the term $(\Theta/Z)$ is a normalizing coefficient that is introduced to remove the effect of unit.

To determine those parameters that significantly affect the steady-state availability, we evaluate sensitivity coefficient for each parameter and then arrange their nonnegative values in decreasing order. The sensitivity ranking obtained through this technique also facilitates in identifying those parameters that have less impact on the availability of the system. Parameters having higher sensitivity coefficient values affect the availability most. On the other hand, the parameters having lower sensitivity coefficient values have least effect on the availability.

Table 3.8 displays the sensitivity ranking of steady-state availability of VCC network for each input parameter by using eqs. (3.10) and (3.11).

In the table the top-ranked parameters are the failure and repair rates of storage pool, input–output interface, API, Walrus, and CLC, respectively. Input–output interface and API both are interfaces that facilitate the user to access the services of vehicular cloud system. On failure of any of these, the VCC services will become disabled for the users which will affect its availability. Also, storage pool and Walrus act as storage systems in which all the data related to VCC services are stored. Hence, their failure will make the VCC system unavailable to the users. CLC controls the complete cloud infrastructure. Therefore, failure of CLC will result in the unavailability of the VCC services for the users. Hence, all the above-discussed parameters are of more importance as compared to other parameters and should be given higher priority as compared to others in improving the availability of the system.

The next set of parameters are related to the inside vehicular environment. Availability of these components is also important so that the user can connect and coordinate with the outside environment. Failure and repair rate of 5G network is also important as it is the only medium that facilitates communication among the users.

In the sensitivity ranking the parameters appearing at the last are related to cloud storage and cloud platform. Due to its parallel structure in the RBD it is expected that they will have lower sensitivity ranking.

**Table 3.8:** Sensitivity ranking based on partial derivatives.

| Sensitivity ranking | Parameter (Θ) | Description | $\|SS_\Theta[A]\|$ |
|---|---|---|---|
| 1 | $\lambda_{storage\_pool}$ | Failure rate-storage pool | 0.00458 |
| 2 | $\mu_{storage\_pool}$ | Repair rate-storage pool | 0.00458 |
| 3 | $\lambda_{I/O}$ | Failure rate-I/O interfaces | 0.002347 |
| 4 | $\mu_{I/O}$ | Repair rate-I/O interfaces | 0.002347 |
| 5 | $\lambda_{API}$ | Failure rate-API | 0.002347 |
| 6 | $\mu_{API}$ | Repair rate-API | 0.002347 |
| 7 | $\lambda_{Walrus}$ | Failure rate-Walrus | 0.001265 |
| 8 | $\mu_{Walrus}$ | Repair rate-Walrus | 0.001265 |
| 9 | $\lambda_{rep\_CLC}$ | Repair rate-CLC | 0.00101 |
| 10 | $\lambda_{sensors}$ | Failure rate-Sensors | 0.000294 |
| 11 | $\mu_{sensors}$ | Repair rate-Sensors | 0.000294 |
| 12 | $\lambda_{GPRS}$ | Failure rate-GPRS | 0.000291 |
| 13 | $\mu_{GPRS}$ | Repair rate-GPRS | 0.000291 |
| 14 | $\lambda_{CU}$ | Failure rate-CU | 0.000284 |
| 15 | $\mu_{CU}$ | Repair rate-CU | 0.000284 |
| 16 | $\lambda_{GPS}$ | Failure rate-GPS | 0.000192 |
| 17 | $\mu_{GPS}$ | Repair rate-GPS | 0.000192 |
| 18 | $\lambda_{5G}$ | Failure rate-5G | 0.000120467 |
| 19 | $\mu_{5G}$ | Repair rate-5G | 0.000120467 |
| 20 | $\lambda_{PS\_server}$ | Failure rate-physical storage server | 0.000019619 |
| 21 | $\mu_{PS\_server}$ | Repair rate-physical storage server | 0.000019619 |
| 22 | $\lambda_{VC\_server}$ | Failure rate-virtual compute server | 0.00001193 |
| 23 | $\mu_{VC\_server}$ | Repair rate-virtual compute server | 0.00001193 |
| 24 | $\lambda_{SC}$ | Failure rate-storage controller | 0.000000107609 |
| 25 | $\mu_{SC}$ | Repair rate-storage controller | 0.000000107609 |
| 26 | $\lambda_{rep\_CC}$ | Failure rate-CC | 0.0000000011979 |
| 27 | $\lambda_{NC}$ | Failure rate-node controller | 0.0000000000005167 |
| 28 | $\mu_{NC}$ | Repair rate-node controller | 0.0000000000005167 |

Figure 3.6 provides a graphical representation of the steady-state availability with respect to the first 15 parameters of Table 3.8. Only first 15 parameters have been presented in Figure 3.6, because for the other parameters the variation in the availability is unnoticeable. It just confirms the conclusion that the low-ranked parameters should get less importance when one is aiming to improve the availability measure.

To further verify the results obtained from partial derivatives, sensitivity analysis through percentage difference technique is performed.

#### 3.5.2.2 Percentage difference technique

In this technique we vary one input parameter from its minimum value to its maximum value. The main advantage of this technique over partial derivative technique is

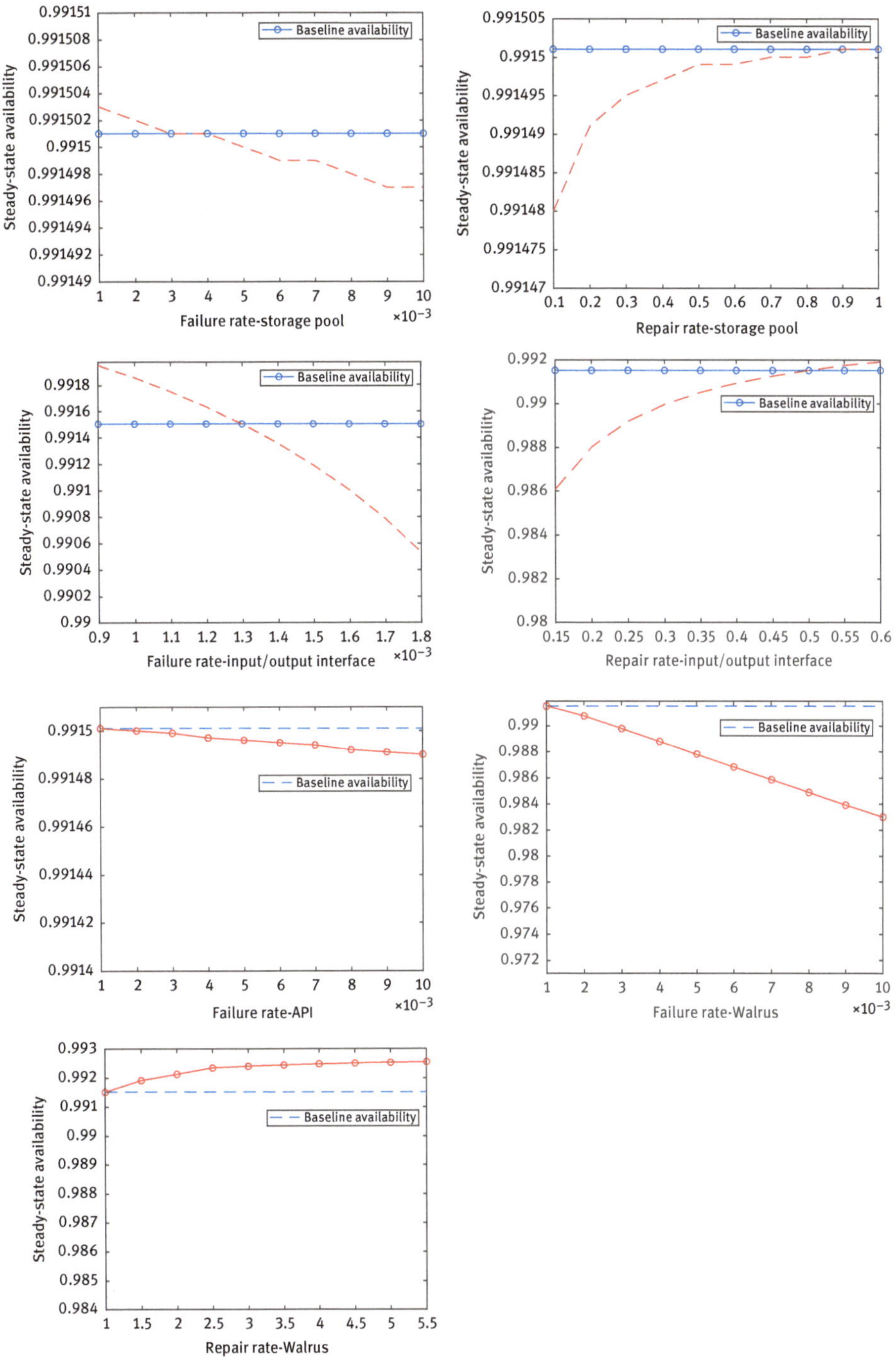


**Figure 3.6:** Graphical representation of variation in steady-state availability.

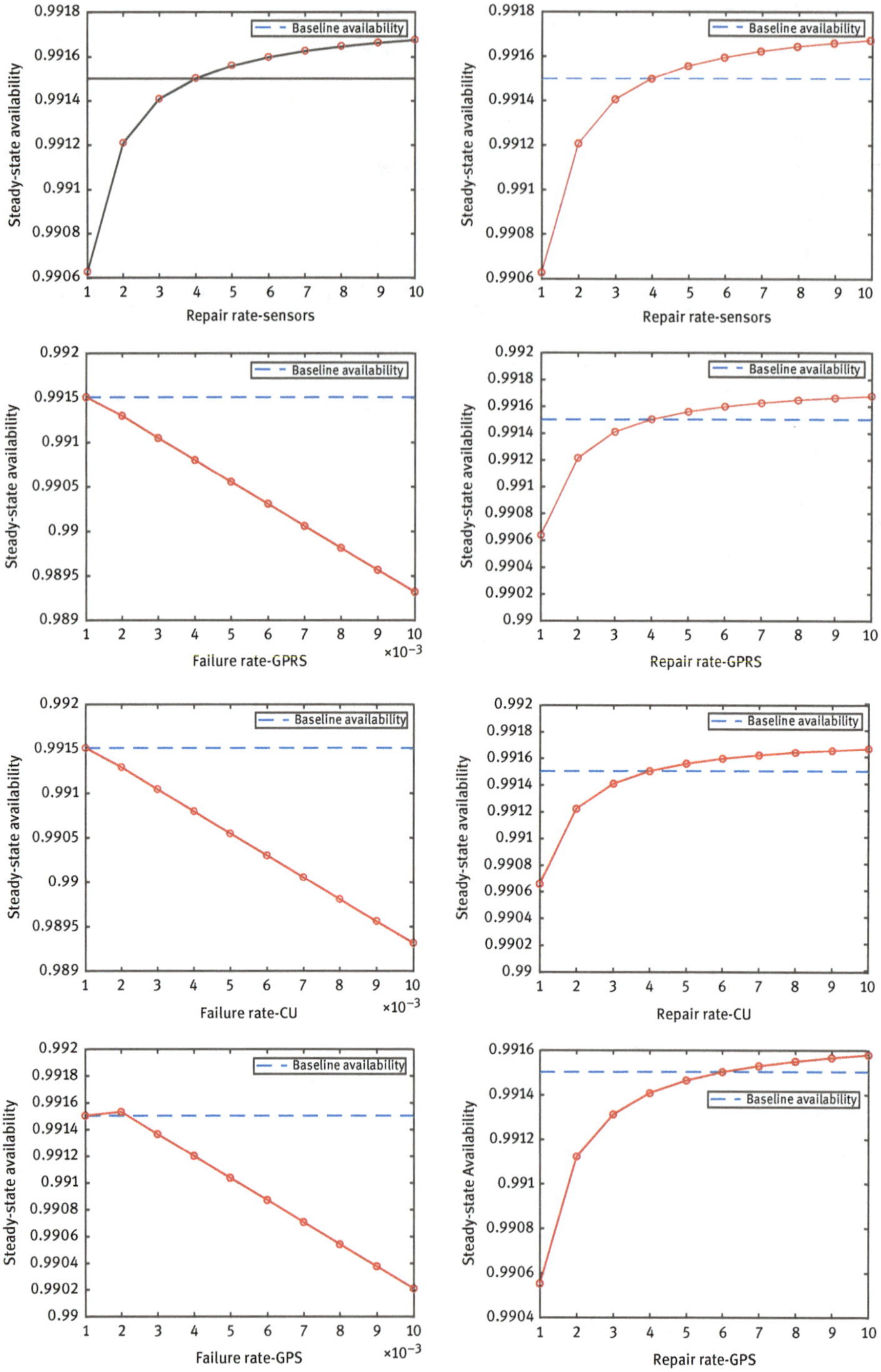


**Figure 3.6** (continued)

that it utilizes the complete range of all possible values of a parameter to evaluate the sensitivities. The mathematical expression for this approach is shown as follows [5]:

$$S_{\theta}(Z) = \frac{\max Z(\theta) - \min Z(\theta)}{\max Z(\theta)} \tag{3.12}$$

where

$$\max Z(\theta) = \max\{Z(\theta_1),\ Z(\theta_2),\ \ldots,\ Z(\theta_n)],$$

and

$$\min Z(\theta) = \min\{Z(\theta_1),\ Z(\theta_2),\ \ldots,\ Z(\theta_n)].$$

where $Z(\theta)$ is the value of the measure $Z$ for the input value $\theta$ and max $Z(\theta)$ and min $Z(\theta)$are the maximum and minimum output values, respectively, obtained by varying the input over its entire range [29].

Similar to the partial derivative technique, we first evaluate the sensitivity coefficient for all the parameters and then arrange them in decreasing order. Table 3.9 represents the sensitivity ranking based on the percentage difference. We have mentioned the ranking of only top 18 parameters. We have used the same range of values for all the parameters that we have used to produce the graphical results. It is observed that almost all the top-ranked parameters obtained from percentage

**Table 3.9:** Sensitivity ranking based on percentage difference.

| Sensitivity ranking | Parameter | \|S(A)\| |
|---|---|---|
| 1 | $\lambda_{SC}$ | 0.025717 |
| 2 | $\lambda_{Walrus}$ | 0.008592 |
| 3 | $\mu_{I/O}$ | 0.005836 |
| 4 | $\mu_{SC}$ | 0.003411 |
| 5 | $\lambda_{CU}$ | 0.002210789 |
| 6 | $\lambda_{Sensors}$ | 0.00220704 |
| 7 | $\lambda_{GPRS}$ | 0.0022004 |
| 8 | $\lambda_{I/O}$ | 0.001423467 |
| 9 | $\lambda_{GPS}$ | 0.001303 |
| 10 | $\mu_{sensors}$ | 0.001058 |
| 11 | $\mu_{GPRS}$ | 0.001045705 |
| 12 | $\mu_{Walrus}$ | 0.001045 |
| 13 | $\mu_{GPS}$ | 0.001034 |
| 14 | $\mu_{CU}$ | 0.001022 |
| 15 | $\mu_{storage\_pool}$ | 2.31971E-05 |
| 16 | $\mu_{API}$ | 1.31114E-05 |
| 17 | $\lambda_{API}$ | 1.10943E-05 |
| 18 | $\lambda_{storage\_pool}$ | 8.06856E-06 |

difference technique are same as those obtained from partial differentiation technique except two parameters ($\lambda_{SC}$ and $\mu_{SC}$).

## 3.6 Conclusion

This chapter analyzes the availability of the VCC network using hierarchical modeling technique. Because of the complex architecture of VCC, distinct models have been developed for each subsystem of VCC, and the results obtained from the submodels are then combined to evaluate the availability of the whole network. RBD and SMP are used to develop the availability models. Sensitivity analysis is also performed to calculate the impact of each parameter on the availability. Two different techniques (partial derivatives technique and percentage difference technique) are used for the sensitivity analysis. The study reveals that the steady-state availability of the system can be significantly improved by just concentrating on a group of parameters that affect the steady-state availability substantially as compared with the other parameters. Based on the two different sensitivity analysis techniques, we get two sensitivity ranking tables providing the list of all those parameters that affect availability the most. It is worth noticing that almost all the parameters are identical in both the tables. According to the tables, the higher ranked parameters are either related to the storage unit of the VCC system or providing access to the users to utilize the VCC facilities. A graphical representation of the steady-state availability with respect to first 15 parameters of the sensitivity ranking table is also provided in the chapter.

**Acknowledgment:** This research work was aided by the University of Delhi. The author (Shivani Gupta) thanks UGC for their financial assistance.

## References

[1] The Car In The Cloud. By Scott Frank, Vice President of Airbiquity. (Accessed March 4, 2019, at https://www.airbiquity.com/blog-events/blog-posts/car-cloud.)

[2] Avizienis, A., Laprie, J.C., Randell, B. et al. (2004). Basic concepts and taxonomy of dependable and secure computing, IEEE Transactions on Dependable and Secure Computing, 1(1), 11–33.

[3] Avizienis, A., Laprie, J.C. and Randell, B. (2001). Fundamental concepts of dependability. University of Newcastle upon Tyne, Computing Science, Apr, 19.

[4] Gupta, V. and Dharmaraja, S. (2011). Semi-Markov modelling of dependability of VoIP network in the presence of resource degradation and security attacks, Reliability Engineering & System Safety, Dec, 1, 96(12), 1627–1636.

[5] Matos, R., Araujo, J., Oliveira, D. et al. (2015). Sensitivity analysis of a hierarchical model of mobile cloud computing, Simulation Modelling Practice and Theory, Jan(50), 1, 151–164.
[6] Dantas, J., Matos, R., Araujo, J. et al. (2012). Models for dependability analysis of cloud computing architectures for eucalyptus platform, International Transactions on Systems Science and Applications, Dec, 8(5), 13–25.
[7] Abuelela, M. and Olariu, S. Taking vanet to the clouds. In Proceedings of the 8th International Conference on Advances in Mobile Computing and Multimedia. ACM 2010, 6–13.
[8] Arif, S., Olariu, S., Wang, J. et al. (2012). Data centre at the Airport: Reasoning about Time-Dependent Parking Lot Occupancy, IEEE Transactions Parallel Distributed Systems., 23(11), 2067–2080.
[9] Olariu, S., Hristov, T. and Yan, G. (2012). The Next Paradigm Shift: From Vehicular Networks to Vehicular Clouds, Mobile Ad Hoc Network-ing: The Cutting-Edge Directions.
[10] Abdelhamid, S., Benkoczi, R. and Hassanein, H.S. (2017). Vehicular clouds: ubiquitous computing on wheels, In Emergent Computation, 435–452.
[11] Gu, L., Zeng, D. and Guo, S. (2013). Vehicular cloud computing: A survey, In 2013 IEEE Globecom Workshops (GC Workshops), Dec, 9, 403–407.
[12] Mekki, T., Jabri, I., Rachedi, A. et al. (2017). Vehicular cloud networks: Challenges, architectures, and future directions, Vehicular Communications, Jul, 1(9), 268–280.
[13] Yan, G., Wen, D., Olariu, S. et al. (2012). Security challenges in vehicular cloud computing, IEEE Transactions on Intelligent Transportation Systems, Sep, 3, 14(1), 284–294.
[14] Al Ridhawi, I., Aloqaily, M., Kantarci, B. et al. (2018). A continuous diversified vehicular cloud service availability framework for smart cities, Computer Networks, Nov, 9(145), 207–218.
[15] Limbasiya, T. and Das, D. (2019). Secure message confirmation scheme based on batch verification in vehicular cloud computing, Physical Communication, Jun, 1(34), 310–320.
[16] Hou, L., Zheng, K., Chatzimisios, P. et al. (2018). A Continuous-Time Markov decision process-based resource allocation scheme in vehicular cloud for mobile video services, Computer Communications, Mar, 1(118), 140–147.
[17] Midya, S., Roy, A., Majumder, K. et al. (2018). Multi-objective optimization technique for resource allocation and task scheduling in vehicular cloud architecture: A hybrid adaptive nature inspired approach, Journal of Network and Computer Applications, Feb, 1(103), 58–84.
[18] Talebifard, P. and Leung, V.C. (2013). Towards a content-centric approach to crowd-sensing in vehicular clouds, Journal of Systems Architecture, Nov, 1, 59(10), 976–984.
[19] Safi, Q.G., Luo, S., Wei, C. et al. (2018). Cloud-based security and privacy-aware information dissemination over ubiquitous VANETs, Computer Standards & Interfaces, Feb, 1(56), 107–115.
[20] Al-Rashed, E., Al-Rousan, M. and Al-Ibrahim, N. (2017). Performance evaluation of wide-spread assignment schemes in a vehicular cloud, Vehicular Communications, Jul, 1(9), 144–153.
[21] Whaiduzzaman, M., Sookhak, M., Gani, A. et al. (2014). A survey on vehicular cloud computing, Journal of Network and Computer applications, Apr, 1(40), 325–344.
[22] Understanding Cloud: Infrastructure-as-a-Service vs. Platform-as-a-Service. (Accessed on March 10, 2019, at https://www.abcservices.com/understanding-cloud-infrastructure-as-a-service-vs-platform-as-a-service/)
[23] Nurmi, D., Wolski, R., Grzegorczyk, C., et al. The eucalyptus open-source cloud-computing system. In Proceedings of the 2009 9th IEEE/ACM International Symposium on Cluster Computing and the Grid 2009 May 18, 124–131.
[24] Matos, R., Dantas, J., Araujo, J. et al. (2017). Redundant eucalyptus private clouds: Availability modelling and sensitivity analysis, Journal of Grid Computing, Mar, 1, 15(1), 1–22.

[25] Dharmaraja, S., Vinayak, R. and Trivedi, K.S. (2016). Reliability and survivability of vehicular ad hoc networks: An analytical approach, Reliability Engineering & System Safety, Sep, 1(153), 28–38.
[26] Cloud Storage. (Accessed on March 20, 2019 at https://en.wikipedia.org/wiki/Cloud_storage)
[27] Trivedi, K.S., Kim, D.S., Roy, A., et al. Dependability and security models. In 2009 7th International Workshop on Design of Reliable Communication Networks 2009, 11–20.
[28] Kulkarni, V.G. (2016). Modelling and analysis of stochastic systems, CRC Press.
[29] Hamby, D.M. (1994). A review of techniques for parameter sensitivity analysis of environmental models, Environmental Monitoring and Assessment., Sep, 1, 32(2), 135–154.

Gabriel Ricardo Pena and Nestor Ruben Barraza

# 4 Increasing failure rate software reliability models for agile projects: a comparative study

**Abstract:** A new software reliability model is proposed. The model is inspired in the Polya stochastic process, which is the asymptotic limit of the Polya urn model for contagion that is well described as a pure birth process. Modeling software reliability with this type of process introduces failure rate functions that depend not only on time but also on the number of failures previously detected. Since the failure rate function of the Polya stochastic process results in a linear-over-time mean number of failures, we propose a new pure birth process with a different failure rate function. This proposal results in a nonlinear-over-time mean number of failures and allows to model not only cases with increasing in time failure rate but also a reliability growth. We consider that failure reports collected under modern software engineering methodologies will require not just reliability growth models but also those which take into account increasing failure rate cases, if they are to be applied at the very first stage of development and testing, when code is constantly added either to fix failures or to accomplish new requirements. We show applications of our proposed model to several datasets and compare the performance with nonhomogeneous Poisson process models.

**Keywords:** software reliability, delayed S-shaped, Goel–Okumoto, logistic, contagion, Agile processes, pure birth process

## 4.1 Introduction

Since software is quite an important component in many technological projects, software quality has became a vital issue in software engineering. It is in that sense that software reliability engineering proposes models that help determine how reliable a delivered software product is or when it will be ready for release. There is an extensive bibliography and reviews on software reliability models. It exceeds the purpose of this work to make even a short list. This study initiates the group of models that can be described by pure birth processes. Pure birth processes are stochastic processes that describe the behavior of a population where individuals can

**Gabriel Ricardo Pena, Nestor Ruben Barraza,** Universidad Nacional de Tres de Febrero, Buenos Aires, Argentina

https://doi.org/10.1515/9783110619058-004

only be born, see for example [1], and hence they can then be used to model the random failure detection process. Many software reliability models are special types of pure birth processes, such as the well known based on nonhomogeneous Poisson processes, first developed in the 1970s by authors such as Musa, Okumoto, and Yamada, which are still a matter of study. Nonhomogeneous Poisson processes are pure birth processes where the mean number of failures is a nonlinear function of time that models a reliability growth. The general point of view under pure birth processes allows us to think about general dependencies of the failure rate not only on time but also on the number of previously detected failures. We point out that pure birth processes have been already used in software reliability, though without the general analysis we make in this work. One of the first software reliability models was proposed in [2], the well-known Jelinski–Moranda model formulated as a birth process is shown in [3] and some recent applications can be found in [4, 5].

Several proposals for the failure (birth) rate that depend on time and on the number of previously detected failures allow us to capture physical mechanisms in either introduced or detected failures. That is the case of contagion. Contagion can be modeled by the Polya urn model, which consists of extracting balls with replacement from an urn and at the same time adding the same number of balls of the same color, rising the probability of extracting a ball of the same color in the next extraction. This model leads to the Polya stochastic process as the limit from discrete to continuous time. The Polya process is a particular case of a pure birth process where the birth (failure) rate depends on time and on the number of individuals (failures) in the population.

A stochastic software failure detection process, modeled by a contagion process, could reveal a contagion phenomena in introduced or detected failures. This fact would be indicative of a sort of interaction, perhaps due to the relations between programmers, software modules, or testers.

A contagious software reliability model could model the software failure stochastic process at the very start of the testing stage, where new code is added to fix failures, or in agile environments, where testing and development phases are simultaneously carried out. We consider that the well-known stochastic processes developed decades ago need to be improved in order to take into account the recent advances in software development engineering as well as testing processes. Despite the fact that some improvements have been made (see, for example [6], where the inclusion of human factors, actual process monitoring, and performability evaluation is considered and a Markovian model is introduced), we consider that such improvements should involve cases in which development and testing are performed simultaneously. For example, at the very start of a software project or when development is performed under modern methodologies such as Agile, which include many innovative practices such as test-driven development (TDD). Then, if we are interested in applying a software reliability model since the beginning of the project, we should consider models that allow an increasing failure rate.

There are no many recent datasets with failure reports collected at the very first stage of testing and development that take into account modern software engineering methodologies. This lack of datasets agrees with the fact that most of the research in software reliability remains in the academy, as shown in recent studies, see for example [7]. Industrial papers are mainly focused on pre- and postrelease studies, where the development process is complete, see for example [8–10].

The analysis of software reliability models as pure birth processes has already been considered in the literature. A general survey and discussion of pure birth processes applied to software reliability can be found in [11]. This type of processes are also applied in the literature as continuous-time Markov chains. Our novel approach is to take advantage of this general formulation to consider more general functional forms of the failure rate and the introduction of contagion as a new concept.

Conversely, simulation of software failures is another concern in software reliability, see for example, [12, 13]. Our general point of view of software reliability models as pure birth processes yields a simulation method through the exponential waiting time. This simulation produces a failure after a random exponential waiting time that depends on the proposed failure rate.

As mentioned earlier, the Polya stochastic process is a pure birth process with a birth rate that is directly proportional to the number of individuals in the population and inversely proportional to the elapsed time, see [1]. Despite the interest in the Polya model in order to fit increasing failure rate cases, it gives as a result a mean number of failures that depends linearly on time. Then, though simulations have demonstrated that this model fits relatively well some decreasing reliability cases, it is useless when we try to fit a mean number of failures curve. Hence, another contagious model that is a slight modification of the Polya process is proposed. We propose a failure rate that depends on time and previously detected failures, obtaining a nonlinear time-dependent mean number of failures. As an interesting result, this nonlinear dependency can model increasing failure rate as well as reliability growth cases where the curve for the first derivative of the mean number decreases.

Considering increasing failure rate cases is not new and comes from the very beginning of the speciality. It was introduced with the S-shaped model in [14]. Our proposal is to consider these cases as coming from a new concept like contagion.

We applied our model to several real datasets analyzed in the literature and compared the results obtained with those obtained after applying some well-known NHPP models such as the Yamada delayed S-shaped, logistic, and Goel–Okumoto models.

We analyze increasing failure rate and reliability growth cases. Our model compares them thus becoming a useful model for several software failure detection stochastic processes.

## 4.2 Modern practices in software reliability

Software development and testing has been evolved since the waterfall model, popular in the 1970s. There are various practices that can be grouped into three categories: traditional, introduced by Agile, and shared. A nonexhaustive list is as follows:

Traditional
- Destructive testing
- Detailed designs/design specifications
- Formal estimation (e.g., COCOMO, FP)
- Formal specification
- Model checking
- Prototyping
- Security testing
- Use case modeling (as requirements engineering practice)

Introduced by Agile
- Iteration/sprint reviews
- Limit working progress (e.g., using Kanban board)
- Onsite customer
- Pair programming
- Refactoring
- Release planning
- Retrospectives
- Scrum of Scrums
- Test-driven development (TDD)
- User stories (as requirements engineering practice)
- Velocity-based planning

Shared
- End to end (system) testing
- Expert/team-based estimation (e.g., Planning Poker)
- Iteration planning

Many of the above-mentioned practices (iteration/sprint reviews, Scrum of Scrums, test-driven development, iteration planning) involve an iteration mechanism that implies a simultaneous development and testing processes. This stage is not a reliability growth one, but contrarily, a stage where we can expect an increasing failure rate stochastic process. Then, the focus is on models that take into account an increasing failure rate first stage and try to predict when the reliability growth starts, see [15] and references therein.

# 4.3 Proposed model

## 4.3.1 Mathematical background

We propose to get the probability of having $r$ failures in a given time $t$ from a pure birth equation.

Pure birth processes describe the behavior of a population where individuals can only be born and are not allowed to die. The probability of having $r$ individuals in the population at a given time $t$, $P_r(t)$, is given by the solution of the following differential equation (4.1):

$$\begin{cases} P'_r(t) = -\lambda_r(t)P_r(t) + \lambda_{r-1}(t)P_{r-1}(t) \\ P'_0(t) = -\lambda_0(t)P_0(t) \end{cases} \tag{4.1}$$

where $\lambda_r(t)$ is the birth (failure) rate, as shown in [1]. The process given in (4.1) is also Markov's, where the number of individuals corresponds to the state of the system, $S_r(t)$. The only transitions allowed (actually, other transitions are infinitesimals of higher order) are $S_r(t) \to S_r(t+dt)$ and $S_r(t) \to S_{r+1}(t+dt)$ with probabilities given by (4.2) and (4.3):

$$P(S_r(t) \to S_{r+1}(t+dt)) = \lambda_r(t)dt \tag{4.2}$$

$$P(S_r(t) \to S_r(t+dt)) = 1 - \lambda_r(t)dt \tag{4.3}$$

The dependency of $\lambda_r(t)$ on $t$ defines the type of the process. If $\lambda_r(t)$ is a function of $t$ or a constant, the process is nonhomogeneous or homogeneous respectively but with independent increments. If $\lambda_r(t)$ is a function of $r$, the process is of dependent increments.

From (4.1), the probability of having no births in a given time interval greater than $t - s$ given the system is at the state $r$ by the time $s$ is given by the well-known exponential waiting time (4.4):

$$P(\text{no births} \in T > t - s) = \exp\left( - \int_s^t \lambda_r(t)dt \right) t \geq s \tag{4.4}$$

The integrate of $\lambda_r(t)$ (4.5) is the mean number of births (failures):

$$\int_s^t \lambda_r(t)dt = \mu(t) - \mu(s) \tag{4.5}$$

The mean number of individuals in a given time (4.6)

$$M(t) = \sum_{r=0}^{+\infty} rP_r(t) \tag{4.6}$$

can be obtained from summing up (4.1) multiplied by $r$, and getting this way a differential equation for the mean (4.7):

$$M'(t) = \sum_{r=1}^{+\infty} \lambda_{r-1} P_{r-1}(t) \tag{4.7}$$

## 4.3.2 Proposed failure rate

Taking into account that in a dynamic project, like those developed under Agile methodologies, new code is constantly added either to fix failures or in order to meet new requirements, we can consider that failures are removed on one hand, and new ones are introduced on the other. Under the hypothesis that new failures are introduced at a rate proportional to the number of failures previously detected (contagion), the proposed failure rate results in a factor that increases proportionally to the previous number of failures and a denominator that increases on time. Then the following mathematical form (4.8) is proposed:

$$\lambda_r(t) = \frac{1}{a}\frac{1 + br}{1 + at} \tag{4.8}$$

If $b$ is zero, the model behaves like the Musa–Okumoto software reliability growth model (see Table 4.1). Equation (4.8) is similar to that of the Polya contagion process given by (4.9) (see [16]):

$$\lambda_r(t) = \rho\frac{r + \gamma}{1 + \rho t} \tag{4.9}$$

With our proposal, we get a quite different mean number of failures as it is explained later. As it was stated earlier, looking at (4.8), our model is a nonhomogeneous Markov with dependent increments process.

**Table 4.1:** NHPP software reliability models.

| Model | $\mu(t)$ |
|---|---|
| Goel–Okumoto | $a(1 - \exp(-bt))$ |
| Musa–Okumoto | $\frac{1}{\theta}\ln(\mu_0\theta t + 1)$ |
| Delayed S-shaped | $a(1 - (1 + bt)\exp(-bt))$ |
| Logistic | $\frac{a}{1 + \exp(-b(t - c))}$ |

Looking at (4.8) the main assumptions of our model can be stated as follows:

- Failures are continuously introduced and removed.
- Code is being constantly added either to fix failures and/or to meet new requirements.
- The new added code introduces failures at a rate proportional to the existing number of failures (contagion).
- Because of the failures fixing process, the failure intensity decreases inversely proportional to the elapsed execution time. This assumption is that of the Musa–Okumoto model.[1]

Using (4.8), the mean number of failures as obtained from (4.7) is obtained by solving the following differential equation:

$$M'(t) = \frac{a}{1+at}(1+bM(t)) \tag{4.10}$$

resulting in the mean value function given by (4.11):

$$M(t) = \frac{1}{b}\left((1+at)^b - 1\right) \tag{4.11}$$

Our proposal, inspired in the Polya contagion process, gives an interesting function of the mean number of failures, resulting in either positive or negative concavities according to whether the $b$ parameter is greater or lower than one, allowing modeling not just increasing failure rate cases but also a reliability growth. The probability of failures can be expressed in an integral form as shown in [17]. Our proposed pure birth process can be simulated in a similar way as it was done for the Polya process in [16] and other pure birth processes as shown in [18], by getting a failure after a randomly generated exponential waiting time.

## 4.4 Nonhomogeneous poisson processes used for comparison

The Poisson process arises from a pure birth process when the birth rate is constant: $\lambda_r(t) = \lambda$. Nonhomogeneous Poisson processes arise when the birth rate is a

1 Actually, in the Musa–Okumoto model, it can be shown that the failure intensity decays exponentially with failures experienced. This factor should be better referred to as Musa–Okumoto like factor.

function of time: $\lambda_r(t) = \lambda(t)$. The probability of having $r$ failures in a time interval $(s,t)$ is given by

$$P((N(t) - N(s)) = r) = \frac{(\mu(t) - \mu(s))^r}{r!} \exp(-(\mu(t) - \mu(s))) \tag{4.12}$$

Different software reliability models based on nonhomogeneous Poisson processes have been proposed in the literature, a short list of them that are used further for comparison is shown in Table 4.1, besides the Musa–Okumoto model mentioned in the previous section. A more extensive list can be found in [19].

## 4.5 Mean time between failures

The mean time between failures (MTBF) is calculated as usual from the density function of the time to the next failure. That density function is obtained as the first derivative of the distribution function $F_T(t) = P(T \le t)$, which corresponds to the complement of the exponential waiting time (4.4). This distribution function is given by

$$F_T(t;r,s) = 1 - \exp\left(-\int_s^t \lambda_r(t)dt\right) \tag{4.13}$$

Thus, by differentiating (4.13), we get the probability density function of the time to the next failure:

$$f_T(t;r,s) = \lambda_r(t)\exp\left(-\int_s^t \lambda_r(t)dt\right) \tag{4.14}$$

By taking the expectation of (4.14) we get the conditional MTBF predicted by our model given $r$ failures were detected by the time $s$ (4.15):

$$\mathrm{MTBF}(r,s) = \frac{1}{a}\frac{1+as}{br}, \quad r = 1, 2, 3, \ldots \tag{4.15}$$

The obtained MTBF depends on two competitive factors: a reliability growth factor depending on time, and a second factor that is inversely proportional to the number of failures, leading to a decrease in the MTBF. Replacing the mean number of failures (4.11) in (4.15) as $r = M(s)$ leads to

$$\mathrm{MTBF}(s) = \frac{1}{a}\frac{1+as}{(1+as)^b - 1} \tag{4.16}$$

By looking at (4.16), we observe that the asymptotic behavior for great values of $s$ is as follows (4.17):

$$\mathrm{MTBF}(s) \to \frac{1}{a}(as)^{b-1} \tag{4.17}$$

We can see again the different behaviors depending on $b$ is greater or lower than one. Both behaviors will be analyzed in the experiments (see Section 4.6).

A failure dataset modeled by our model could reveal a contagion process either during the development or testing phases. This contagion phenomenon must be analyzed in each particular case and could be due to a sort of interaction between programmers or testers, some characteristics of a particular piece of code, using a given module several times, and so on.

For NHPP, the MTBF is calculated with the standard formulation, from the density function of the time to failure number $k$ (see [19] Eq. (6.78)). Let $T_k$ denote the time until the $k$-th failure; then, its expected value is given by (4.18):

$$E[T_k] = \frac{\int_0^{+\infty} z\lambda(z)[\mu(z)]^{k-1}\exp(-\mu(z))dz}{\int_0^a z^{k-1}\exp(-z)dz}, \quad a = \lim_{t\to+\infty}\mu(t) \tag{4.18}$$

Finally, by defining $X_k = T_k - T_{k-1}$, the MTBF is obtained as follows:

$$E[X_k] = E[T_k] - E[T_{k-1}], \quad k = 1, 2, 3, \ldots \tag{4.19}$$

## 4.6 Experiments

In this section, we analyze the application of three well-known models based on nonhomogeneous Poisson processes (the Goel–Okumoto model, the Yamada delayed S-shaped model, and the logistic model), and our proposed contagion model to several failure datasets comparing the results. In order to evaluate the goodness of fit, we calculate the predictive ratio risk (PRR) and the Akaike information criterion (AIC) for every model. Two different parameter estimation procedures were performed for each of the three NHPP models: least-squares over the mean number of failures curve, and maximum likelihood (taking the least-squares fitted parameters as the initial approximation to solve the maximum likelihood equations). Due to the lack of a closed formula for the failure time pdf (see [17]) we are unable to perform a maximum likelihood estimation on the contagion model nor calculate the exact MTBFs; however, we use (4.16) in order to compute conditional MTBFs and MTTFs based on the datasets. In order to evaluate the predictive validity of each model, we calculate the PRR on the whole dataset through estimating the parameters by short stages. The MTBF curves for these types of projects exhibit a first increasing failure rate stage, what resembles the well-known hardware reliability bathtub curve, with an infant mortality first stage. This is an important characteristic we want to remark for these type of projects developed under modern software and testing methodologies. The integral in the numerator of (4.18) must be calculated by the saddle point method for values of $k$ greater than 100 (see [20]).

### 4.6.1 The NTDS project

This is a classic dataset that has been extensively analyzed in the literature, see for example [21] or [19]. The data come from the Naval Fleet Computer Programming Center and corresponds to the development of the Naval Tactical Data System's software core. We considered only the first 26 failures, which were found during the production phase (for a total of 250 days).

The development of this project was made under traditional methodologies, that is, the waterfall model. Besides the typical reliability growth stage obtained for this type of development methodology, the dataset exhibits an S-shaped curve, with an increasing failure rate first stage. The four processes we analyzed were able to fit this dataset smoothly. Table 4.2 shows the estimated parameters. Figure 4.1 shows the mean value curve for every model, obtained by the least-squares parameters.

**Table 4.2:** NTDS project, estimated parameters.

| Parameter | Goel–Okumoto | | Delayed S-shaped | | Logistic | | Our model |
|---|---|---|---|---|---|---|---|
| | LS | ML | LS | ML | LS | ML | LS |
| $a$ | 33.5994 | 33.9935 | 26.7155 | 27.4915 | 24.6114 | 24.6114 | 0.3799 |
| $b$ | 0.0063 | 0.0058 | 0.0212 | 0.0186 | 0.0413 | 0.0413 | 0.6450 |
| $c$ | – | – | – | – | 76.4858 | 76.4858 | – |

Since the NTDS was developed under the traditional paradigm (waterfall lifecycle), it is reasonable to expect convex curves due to reliability growth. The real data does not exhibit a pure reliability growth behavior but two well-distinguished stages, giving as a result an S-shaped curve. The DS and logistic models fit well the first stage and predict a reliability growth. The Goel–Okumoto model fits the best the late stage of the project (large $t$ and $n$). Table 4.3 shows the PRR and the AIC for each of the four models.

The best results for this project are obtained for the logistic model, which predicts very accurately the number of failures at every stage of the project.

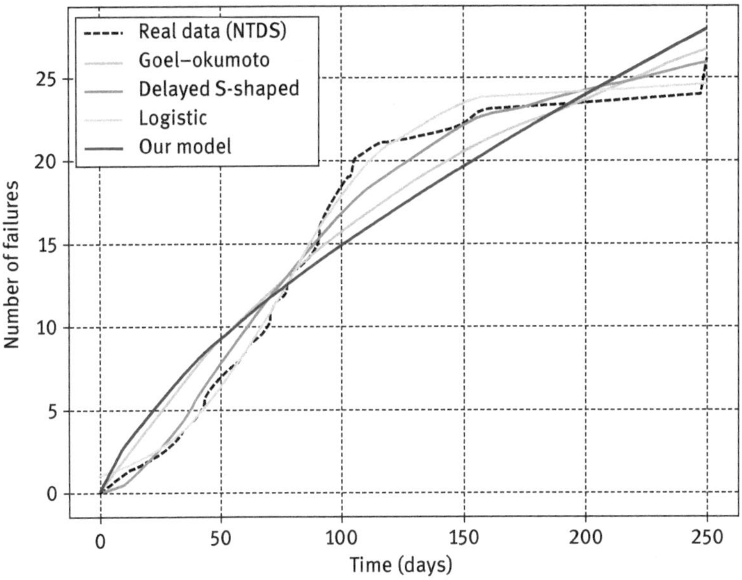


**Figure 4.1:** NTDS data, mean value curves.

**Table 4.3:** NTDS project, fit metrics.

| | Goel–Okumoto | Delayed S-shaped | Logistic | Our model |
|---|---|---|---|---|
| PRR (LS) | 1.5072 | 2.0448 | 0.1811 | 1.9854 |
| PRR (ML) | 1.4050 | 3.9905 | 0.1811 | – |
| AIC | 169.3803 | 165.8360 | 180.5159 | – |

This can be observed in both the fit metrics and the curves. The extremely low PRR values suggest that the fit is especially good on the latest part of the curve. It is also remarkable that the least-squares and maximum likelihood methods give similar results. On the other hand, PRR is significantly higher for the DS model, due to an underestimation of the number of failures at the beginning. Our model behaves better than the DS, but is still outperformed by the other two. Table 4.4 also shows calculations of the PRR over time.

Figure 4.2 shows the (interpolated) MTBF curves for every model (including the conditional MTBF for our proposed model). The Goel–Okumoto model predicts increasing MTBFs due to reliability growth. The other models, however, show more ups and downs. The logistic model predicts a decrease in the MTBFs as $n$ grows, and it has the lowest MTBF value of all four models for $n = 26$; hence, it may not be predicting a reliability growth. Finally, our model gives a slowly increasing MTBF, since it corresponds to a reliability growth model when its $b$ parameter is lower than one.

**Table 4.4:** NTDS project, PRR over time.

| Day | Goel–Okumoto | Delayed S-shaped | Logistic | Our model |
|---|---|---|---|---|
| 70 | N/A | 3.5465 | 9.9849 | 1.9522 |
| 105 | N/A | 4.5698 | 1.0747 | 1.6607 |
| 250 | 1.5072 | 2.0448 | 0.1811 | 1.9854 |

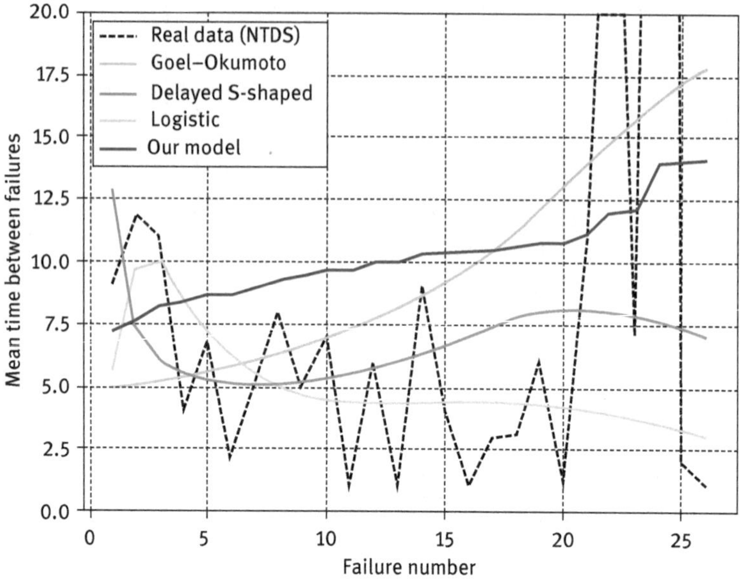


**Figure 4.2:** NTDS data, MTBF curves.

## 4.6.2 The mixed waterfall and agile project

This dataset corresponds to a client–server system developed between 2001 and 2005, which includes around 250 C language kLoC. The project involves both the server and remote terminals applications, linked by a X25/IP WAN. The code also includes a UI for the operation and a relational database managing system.

Development and testing were performed under a combined waterfall/agile methodology, and they both continued after the release (which happened around 110 days after the beginning of the failure report). The report shows a total of 886 failures found within 209 days and registered on the "failures per day" format. Table 4.5 shows the estimated parameters for the four models analyzed; Figure 4.3 shows the mean value curves fitted by the least-squares method.

**Table 4.5:** Mixed waterfall–agile project, estimated parameters.

| Parameter | Goel–Okumoto | | Delayed S-shaped | | Logistic | | Our model |
|---|---|---|---|---|---|---|---|
| | LS | ML | LS | ML | LS | ML | LS |
| *a* | 1416.9139 | 1373.2903 | 893.6389 | 976.8315 | 835.4106 | 835.5226 | 21.0424 |
| *b* | 0.0048 | 0.0050 | 0.0218 | 0.0190 | 0.0307 | 0.0229 | 0.7869 |
| *c* | – | – | – | – | 75.8011 | 88.9528 | – |

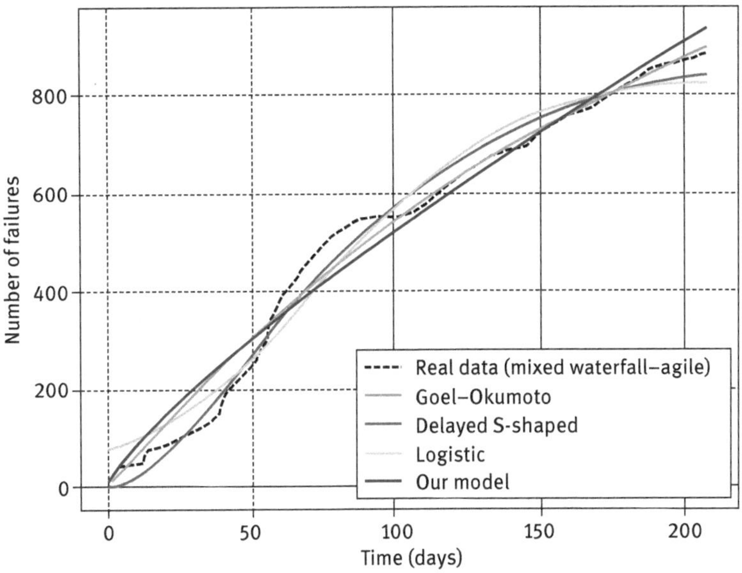


**Figure 4.3:** Mixed waterfall–agile data, mean value curves.

A look at the curve of the real data clearly displays an increasing failure rate stage until the day 60. Then, up to the day 110, the data exhibits the reliability growth behavior. From day 110 to the end of the report, the curve is almost linear, which indicates an almost constant failure rate. Figure 4.3 indicates that the Goel–Okumoto model follows almost perfectly the constant failure rate stage, but is not able to predict the project state at the start of the project, as expected. On the contrary, DS model curve performs a reasonable fit on the whole dataset. Neither of the models show a considerable difference between the least-squares and the maximum likelihood estimations, as seen in Table 4.5. The PRR metrics for the four models are depicted in Table 4.6. We observe large values for the DS model produced by the model underestimation of the early days failure numbers;

**Table 4.6:** Mixed waterfall–agile project, fit metrics.

| | Goel–Okumoto | Delayed S-shaped | Logistic | Our model |
|---|---|---|---|---|
| PRR (LS) | 9.0157 | 2371.3295 | 7.0011 | 7.7002 |
| PRR (ML) | 9.0370 | 3414.6162 | 13.9193 | – |
| AIC | 1291.6985 | 1414.9867 | 1408.1298 | – |

on the other hand, the logistic and our proposed model give the best performance metric in order to adjust the whole dataset. Analysis of the PRR over time is shown in Table 4.7.

**Table 4.7:** Mixed waterfall–agile project, PRR over time.

| Day | Goel–Okumoto | Delayed S-Shaped | Logistic | Our model |
|---|---|---|---|---|
| 50 | N/A | 1,332.0000 | N/A | 39.5233 |
| 60 | N/A | 3,327.0000 | N/A | 199.5184 |
| 70 | N/A | 4,576.0000 | N/A | 426.4233 |
| 90 | N/A | 2,895.0000 | 7.1626 | 152.9020 |

Predicted MTBFs shown in Figure 4.4 for the first 100 days of the dataset are quite different for the models analyzed. Goel–Okumoto model predicts an almost constant MTBF, while DS model shows an unusually high MTBF on the early stages that decays almost exponentially and then becomes almost linear (and still decreasing). Logistic model predicts the exact opposite: an almost constant but slightly increasing MTBF in the early stages that suddenly increases exponentially; this implies that the MTTF also has exponential behavior. Our model predicts a MTBF that grows almost linearly and gradually.

### 4.6.3 The agile #1 project

The dataset we named Agile #1 belongs to a noncritical real-time system developed under Agile paradigm in 2017, consisting in 47 modules with 26,729 Java language and 9,465 XML lines of code. The entire dataset was obtained with the project still in development, with testing performed at the same time when a new code was introduced; an increasing failure rate is expected due to this fact. Since no reliability growth phase is present in the data, the Goel–Okumoto model fit could not be

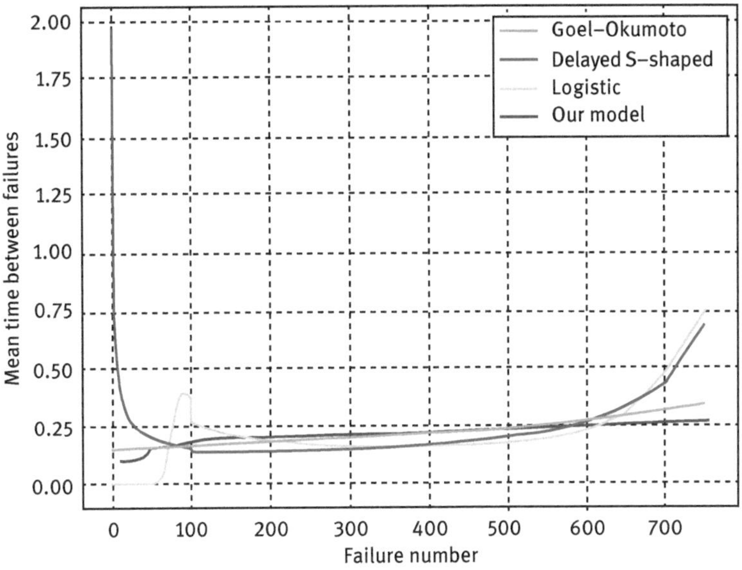


**Figure 4.4:** Mixed waterfall–agile data, MTBF curves.

performed. Table 4.8 shows the obtained parameters for DS, logistic and contagion models, and Figure 4.5 depicts the curves. Again, it is observed that LS and ML estimates are exactly the same for the logistic model, as already shown in Section 4.6.1.

**Table 4.8:** Agile #1 project, estimated parameters.

| Parameter | Delayed S-shaped | | Logistic | | Our model |
|---|---|---|---|---|---|
| | LS | ML | LS | ML | LS |
| *a* | 72.4203 | 63.9117 | 32.3171 | 32.3171 | 0.0239 |
| *b* | 0.0031 | 0.0031 | 0.0117 | 0.0117 | 1.5923 |
| *c* | – | – | 280.7373 | 280.7373 | – |

Goodness-of-fit metrics (Table 4.9) show that the logistic model outperforms the DS significantly. This is true even if the AIC takes a slightly higher value because of the complexity penalty. Our model behaves even better than the logistic. Table 4.10 also shows the PRR calculations over time, obtained by the least-squares estimates.

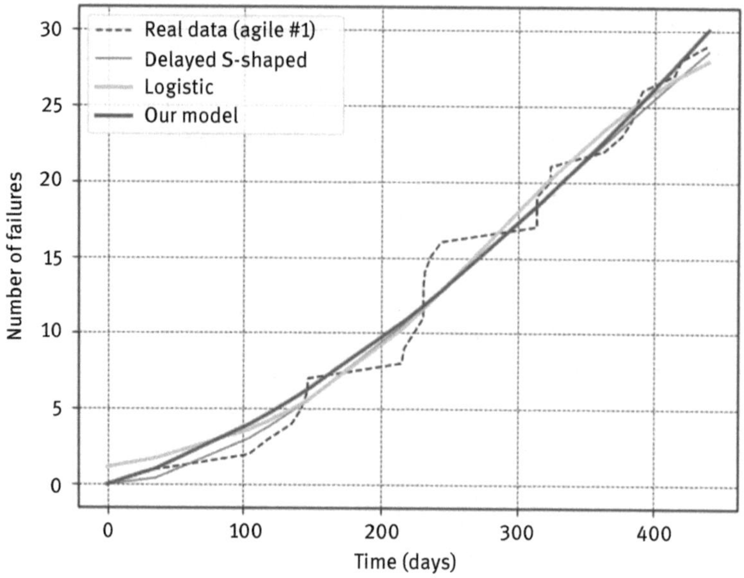


**Figure 4.5:** Agile #1 data, mean value curves.

**Table 4.9:** Agile #1 project, fit metrics.

| | Delayed S-shaped | Logistic | Our model |
|---|---|---|---|
| PRR (LS) | 2.3023 | 0.8704 | 0.7909 |
| PRR (ML) | 3.8577 | 0.8704 | – |
| AIC | 213.9561 | 216.8636 | – |

**Table 4.10:** Agile #1 project, PRR over time.

| Day | Delayed S-shaped | Logistic | Our model |
|---|---|---|---|
| 217 | 6.8003 | 184.3766 | 1.9767 |
| 314 | 2.1844 | N/A | 0.7520 |
| 439 | 2.3023 | 0.8704 | 0.7909 |

The four MTBF curves, including the conditional MTBF curve for the contagion model, are depicted in Figure 4.6. All the models predict similar MTBFs for the most part of the dataset, the logistic model is the only one that behaves differently over the early stages. A significant decreasing on the MTBF is observed as the failure number increases.

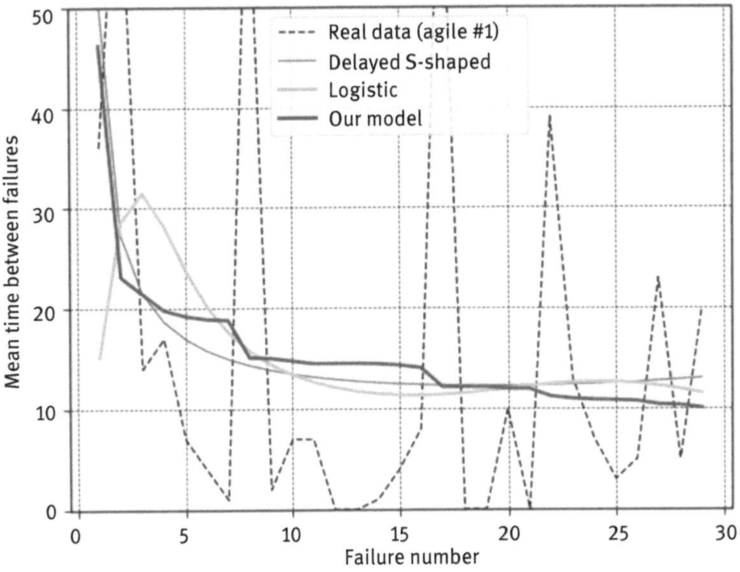


**Figure 4.6:** Agile #1 data, MTBF curves.

### 4.6.4 The agile #2 project

Agile #2 dataset was analyzed in [22], it corresponds to a browser-based testing development tool for web apps, which was written in JavaScript, XML, HTML, and CSS languages and developed under agile methodologies. The failure report includes post-release data from after 11 releases, and the reliability growth stage has not been reached. Delayed S-shaped, logistic, and contagion model estimates are shown in Table 4.11, and its least-squares curves in Figure 4.7. Maximum likelihood fit for logistic model could not be performed due to convergence issues.

**Table 4.11:** Agile #2 project, estimated parameters.

| Parameter | Delayed S-shaped | | Logistic | | Our model |
|---|---|---|---|---|---|
| | LS | ML | LS | ML | LS |
| $a$ | 370.3103 | 329.0763 | 309.2426 | N/A | 0.2221 |
| $b$ | 0.0025 | 0.0030 | 0.0047 | N/A | 1.0479 |
| $c$ | – | – | 580.9822 | N/A | – |

Fit metrics are displayed in Table 4.12. The DS model does not fit well, showing a PRR 20 times larger than the logistic's one. The curve shown in Figure 4.7 also

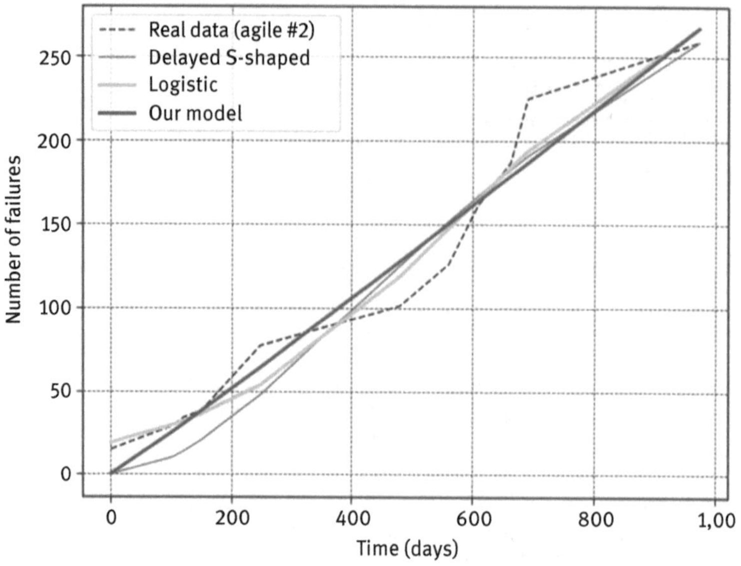


**Figure 4.7:** Agile #2 data, mean value curves.

**Table 4.12:** Agile #2 project, fit metrics.

| | Delayed S-shaped | Logistic | Our model |
|---|---|---|---|
| PRR (LS) | 6.1931 | 0.2571 | 0.1897 |
| PRR (ML) | 3.2098 | N/A | – |
| AIC | 3399.6918 | N/A | – |

shows a poor fit. Our model's PRR is slightly better than the logistic, proving to be the best choice (at least in the PRR sense) for the two agile projects analyzed here. PRR over time results are shown in Table 4.13.

MTBF curves are shown in Figure 4.8. As in the mixed waterfall–agile dataset, the observed time between failures was not registered. Moreover, this report consists on grouped data over nonequal intervals, which means no information is available about each day's failure population; in consequence, formula (4.16) cannot be applied for each day, but only for those whose failure population was registered. This can be clearly seen on the contagion model curve, which consists on multiple linear segments that do not cover the entire time axis. The behavior observed for the predicted MTBFs on the agile #1 project persists here, with a rapidly decreasing DS and contagion curves, and a logistic one that also has a local maximum on the early time stages.

**Table 4.13:** Agile #2 project, PRR over time.

| Day | Delayed S-shaped | Logistic | Our model |
|---|---|---|---|
| 151 | 55.8730 | 15.5480 | 62.0720 |
| 248 | 1.8276 | N/A | 0.5759 |
| 480 | 3.9467 | 4.5307 | 0.9381 |
| 691 | 8.9380 | N/A | 0.5538 |
| 974 | 6.1931 | 0.2571 | 0.1897 |

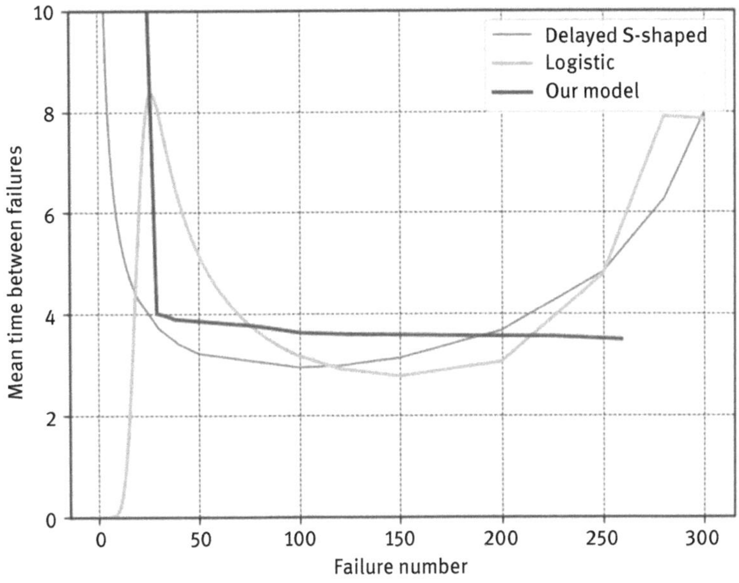


**Figure 4.8:** Agile #2 data, MTBF curves.

## 4.7 Conclusion

A new software reliability model was proposed. The proposed model is a special case of nonhomogeneous Markov pure birth process with a failure rate that depends on the number of failures previously detected and on the elapsed time. The proposed failure rate includes the introduction of new failures as well as the failure removal, as it happens when new code is added to fix failures on the one hand, and to meet new requirements on the other. Our model can be applied both to increasing failure rate and to reliability growth cases. Increasing failure rate cases may arise when software reliability models are applied at the very first stage of development and testing, especially when they are performed under modern software engineering methodologies, such as agile or test-driven development, where testing and development are simultaneously performed. Failure reports that can be modeled by our model could reveal the presence of a contagion phenomena, a sort of interaction between programmers, testers, or pieces of reused code. Applications of our model were compared with the Yamada delayed S-shaped and logistic models in the increasing failure rate stage and with the Goel–Okumoto software reliability growth model when the failure rate decreases. Results of applications show that our model compares well under the PRR (predictive ratio risk) and outperforms the other models in modern projects developed under agile methodologies.

**Acknowledgment:** This work was supported by Universidad Nacional de Tres de Febrero under grant no. 32/473 A.

# Bibliography

[1] Feller, W. (1968). An Introduction to Probability Theory and Its Applications, Vol. 1, 3rd, Wiley, Jan.
[2] Kremer, W. (1983). Birth-death and bug counting, IEEE Transactions on Reliability, R-32(1), 37–47. April.
[3] Jelinski, Z. and Moranda, P. (1972). Software reliability research, Statistical Computer Performance Evaluation, Freiberger, W., Ed, Academic Press, New York, 465–484.
[4] Barraza, N.R., "A parametric empirical bayes model to predict software reliability growth," Procedia Computer Science, vol. 62, pp. 360–369, 2015, proceedings of the 2015 International Conference on Soft Computing and Software Engineering (SCSE'15). [Online]. Available: http://www.sciencedirect.com/science/article/pii/S187705091502551X
[5] Barraza, N.R., "A New Homogeneous Pure Birth Process Based Software Reliability Model," in Proceedings of the 38th International Conference on Software Engineering Companion, ser. ICSE '16. New York, NY, USA: ACM, 2016, pp. 710–712. [Online]. Available: http://doi.acm.org/10.1145/2889160.2892645
[6] Yamada, S., Software Reliability Modeling: Fundamentals and Applications. Springer, 2013. [Online]. Available: https://books.google.com.ar/books?id=T-wlswEACAAJ
[7] Febrero, F., Calero, C. and Moraga, M.A. (2014). A systematic mapping study of software reliability modeling, Information and Software Technology, 56(8), 839–849. [Online]. Available:. http://www.sciencedirect.com/science/article/pii/S0950584914000676.
[8] Rotella, P., Chulani, S. and Goyal, D., "Predicting field reliability," in 2015 IEEE/ACM 3rd International Workshop on Release Engineering, May 2015, pp. 12–15.
[9] Rotella, P. and Chulani, S., "Predicting release reliability," in 2017 IEEE International Conference on Software Quallity, Reliability and Security (QRS2017), Jul 2017, pp. 39–46.
[10] Jalote, P. and Murphy, B., "Reliability growth in software products," in 15th International Symposium on Software Reliability Engineering, Nov 2004, pp. 47–53.
[11] Okamura, H. and Dohi, T., "Unification of software reliability models using markovian arrival processes," in 17th IEEE Pacific Rim International Symposium on Dependable Computing,

PRDC 2011, Pasadena, CA, USA, December 12–14, 2011, Alkalai, L., Tsai, T. and Yoneda, T., Eds. IEEE Computer Society, 2011, pp. 20–27. [Online]. Available: https://doi.org/10.1109/PRDC.2011.12

[12] Gokhale, S.S., Lyu, M.R. and Trivedi, K.S. (2006). Incorporating fault debugging activities into software reliability models: a simulation approach, IEEE Transactions on Reliability, 55(2), 281–292. June.

[13] Lin, C.-T. and Li, Y.-F. (2014). Rate-based queueing simulation model of open source software debugging activities, Software Engineering, IEEE Transactions on, 40(11), 1075–1099. Nov.

[14] Yamada, S., Ohba, M. and Osaki, S. (1983). S-shaped reliability growth modeling for software error detection, IEEE Transactions on Reliability, R-32(5), 475–484. Dec.

[15] Barraza, N.R. (2019). Software reliability modeling for dynamic development environments, Anand, A. and Ram, M., editors, Recent Advancements in Software Reliability Assurance, Advances in Mathematics and Engineering, Chapter 3, CRC Press, pages 29–37.

[16] Barraza, N.R., "Software reliability modeled on contagion," in 2016 IEEE International Symposium on Software Reliability Engineering Workshops (ISSREW), Oct 2016, pp. 49–50.

[17] Sendova, K.P. and Minkova, L.D. (2019). Introducing the nonhomogeneous compound-birth process, Stochastics, 0(0), 1–19.

[18] Barraza, N.R., "Mining bugzilla datasets with new increasing failure rate software reliability models," in XLIII Latin American Computing Conference, CLEI 2017, Cordoba, Argentina, September 4–8, 2017. [Online]. Available: http://www.clei2017-46jaiio.sadio.org.ar/sites/default/files/Mem/SLISW/slisw-04.pdf

[19] Pham, H. (2010). System Software Reliability, 1st, Springer Publishing Company, Incorporated.

[20] Mathews, J. and Walker, R., Mathematical Methods of Physics, ser. Addison-Wesley World Student Series. W. A. Benjamin, 1970. [Online]. Available: https://books.google.com.ar/books?id=1iHvAAAAMAAJ

[21] Hossain, S.A. and Dahiya, R.C. (1993). Estimating the Parameters of a Non-homogeneous Poisson-Process Model for Software Reliability, IEEE Transactions on Reliability, 42(N. 4). December.

[22] Mohamad, S. and McBride,, "Open source, agile and reliability measures," in Software Quality Engineering: Proceedings of the CONQUEST 2009, Jan 2009, pp. 103–117.

Yoshinobu Tamura and Shigeru Yamada

# 5 Fault big data analysis based on effort prediction models and AI for open-source project

**Abstract:** A huge number of open-source software (OSS) is embedded in various software systems and services. In particular, the software systems become increasingly complicated by using many OSS components. Many OSS are embedded in several areas because of the standardization, quick delivery, cost reduction, and so on. Also, various software reliability growth models have been structured in the past.

In this chapter, we discuss several effort prediction models based on the jump diffusion and Wiener processes considering several external factors of OSS project. Then, the method of maximum likelihood, deep learning, and genetic algorithm is used as the parameter estimation method for stochastic differential equation and jump diffusion process models.

Moreover, several numerical examples based on the fault big data in actual OSS projects are shown by using the effort prediction models discussed in this chapter. Finally, the results of parameter estimation in effort prediction models are presented. Furthermore, the method of parameter estimation will be useful to assess the quality and reliability of OSS developed under the open-source project.

**Keywords:** Fault big data, deep learning, effort prediction model, open-source project

## 5.1 Introduction

Recently, various open-source software (OSS) is used in many software systems and services. Also, many software systems have been developed by using several OSS components because of the quick delivery, standardization, cost reduction, and so on. Then, various software reliability growth models for OSS system have been proposed in the past [1–3].

This chapter discusses several effort prediction models considering the Wiener process and AI in order to comprehend several external factors of OSS project. In particular, we discuss the jump diffusion process model. Then, the method of genetic

**Yoshinobu Tamura,** Tokyo City University, Tokyo, Japan
**Shigeru Yamada,** Tottori University, Tottorri, Japan

https://doi.org/10.1515/9783110619058-005

algorithm (GA), deep learning, and maximum likelihood is used as the parameter estimation method based on AI for stochastic differential equation (SDE) models.

Moreover, several numerical examples based on the actual fault big data on the projects of OSS are presented by using the effort prediction models proposed in this chapter. Then, the numerical illustrations of parameter estimation based on AI are discussed in this chapter. Finally, we prove that the proposed effort prediction models will be useful to predict software effort for the quality and reliability of OSS developed under the open source project.

## 5.2 Flexible effort prediction model based on jump diffusion processes

We discuss the modeling of jump diffusion processes to control the OSS maintenance effort during the operation. Let $Z(t)$ be the cumulative OSS maintenance effort expenditures up to time $t$ $(t \geq 0)$ in the operation. $Z(t)$ are the real values continuously. Then, $Z(t)$ gradually increases as the progress of OSS operation, because the maintenance effort is recorded in the OSS operation. By using the modeling technique of classical software reliability modelling [4–7], the following equation considering the OSS maintenance effort is given as follows:

$$\frac{dZ(t)}{dt} = \beta(t)\{\alpha - Z(t)\}, \tag{5.1}$$

where $\beta(t)$ is the effort expenditure rate of OSS at time $t$. $\alpha$ represents the estimated maintenance effort of OSS expended in the period of specified version.

Considering Brownian motion, eq. (5.1) is represented to the following SDE [8, 9]:

$$\frac{dZ(t)}{dt} = \{\beta(t) + \sigma\nu(t)\}\{\alpha - Z(t)\}, \tag{5.2}$$

where $\sigma$ is added as a positive value meaning a level of the irregular continuous fluctuation, and $\nu(t)$ a standardized Gaussian white noise due to development environment. Then, eq. (5.2) is extended to the following SDE of an Itô type:

$$dZ(t) = \left\{\beta(t) - \tfrac{1}{2}\sigma^2\right\}\{\alpha - Z(t)\}dt + \sigma\{\alpha - Z(t)\}dw(t), \tag{5.3}$$

where $w(t)$ means one-dimensional Wiener process. $w(t)$ can be represented as the white noise $\nu(t)$.

The jump term is embedded to the SDE models of eq. (5.3) by considering the unexpected irregular situation at time $t$ by many external complicated project factors. The process of jump diffusion is obtained as follows:

$$dZ_j(t) = \left\{\beta(t) - \tfrac{1}{2}\sigma^2\right\}\left\{\alpha - Z_j(t)\right\}dt + \sigma\left\{\alpha - \Psi_j(t)\right\}dw(t) + d\left\{\sum_{i=1}^{Y_t(\lambda)} (V_i - 1)\right\}, \tag{5.4}$$

where a Poisson point process with frequency $\lambda$ at time $t$ is represented as $Y_t(\lambda)$, that is, the number of jumps, and $\lambda$ the rate of jump. $V_i$ means the range of $i$-th jump. We assume that $w(t)$, $Y_t(\lambda)$, and $V_i$ are independent mutually. Moreover, the increasing rates of OSS maintenance effort for $\beta(t)$ are shown as

$$\int_0^t \beta(s)ds \doteq \frac{\frac{dR_*(t)}{dt}}{\alpha - R_*(t)}, \tag{5.5}$$

$$R_e(t) = a\left(1 - e^{-bt}\right), \tag{5.6}$$

$$R_s(t) = a\{1 - (1 + bt)e^{-bt}\}. \tag{5.7}$$

In this chapter, $\beta(t)$ is assumed the mean value functions in eqs. (5.6) and (5.7) from NHPP (nonhomogeneous Poisson process) models as the OSS effort expenditure function of our model, where $a \doteq \alpha$ is the expected cumulative number of latent faults, and $b \doteq \beta$ the detection rate per fault in terms of software reliability growth models.

Based on It ô's formula [10], $Z_{j*}(t)$ in eq. (5.4) can be derived as

$$Z_{je}(t) = \alpha\left[1 - \exp\left\{-\beta t - \sigma w(t) - \sum_{i=1}^{Y_t(\lambda)} \log V_i\right\}\right], \tag{5.8}$$

$$Z_{js}(t) = \alpha\left[1 - (1 + \beta t)\exp\left\{-\beta t - \sigma v(t) - \sum_{i=1}^{Y_t(\lambda)} \log V_i\right\}\right]. \tag{5.9}$$

Moreover, we extend the existing jump diffusion process model obtained from eq. (5.4) to the following time-delay jump diffusion processes:

In case of $(t \ge 0)$:

$$dZ_{fj}(t) = \left\{\beta(t) - \frac{1}{2}\sigma^2\right\}\{\alpha - Z_{fj}(t)\}dt + \sigma\{\alpha - Z_{fj}(t)\}dw(t) + d\left\{\sum_{i=0}^{Y_t(\lambda_1)} (V_i^1 - 1)\right\}. \tag{5.10}$$

In case of $(t \ge 0,\ t' \ge t_1)$:

$$\begin{aligned} dZ_{fj}(t) = {} & \{\beta(t) - \tfrac{1}{2}\sigma^2\}\{\alpha - Z_{fj}(t)\}dt + \sigma\{\alpha - Z_{fj}(t)\}dw(t) \\ & + d\left\{\sum_{i=0}^{Y_t(\lambda_1)} (V_i^1 - 1)\right\} + d\left\{\sum_{i=0}^{Y_{t'}(\lambda_2)} (V_i^2 - 1)\right\}, \end{aligned} \tag{5.11}$$

where $Y_t(\lambda_1)$ and $Y_{t'}(\lambda_2)$ are Poisson point processes with parameter $\lambda_1$ and $\lambda_2$ at each operation time $(t \ge 0)$ and $(t' \ge t_1)$, respectively. Moreover, $V_i^1$ and $V_i^2$ are $i$th jump ranges in each operation time $(t \ge 0)$ and $(t' \ge t_1)$, respectively. We assume that $Y_t(\lambda_1)$, $Y_t(\lambda_2)$, $V_i^1$, and $V_i^2$ are mutually independent in this paper.

From It ô's formula [10], the solution of eqs. (5.10) and (5.11) can be obtained as follows:

In case of $(t \geq 0)$:

$$Z_{fje}(t) = \alpha\left[1 - \exp\left\{-\beta t - \sigma w(t) - \sum_{i=1}^{Y_t(\lambda_1)} \log V_i^1\right\}\right], \tag{5.12}$$

$$Z_{fjs}(t) = \alpha\left[1 - (1+\beta t)\cdot\exp\left\{-\beta t - \sigma w(t) - \sum_{i=1}^{Y_t(\lambda_1)} \log V_i^1\right\}\right]. \tag{5.13}$$

In case of $(t \geq 0,\ t' \geq t_1)$:

$$Z_{fje}(t) = \alpha\left[1 - \exp\left\{-\beta t - \sigma w(t) - \sum_{i=1}^{Y_t(\lambda_1)} \log V_i^1 - \sum_{i=1}^{Y_{t'}(\lambda_2)} \log V_i^2\right\}\right], \tag{5.14}$$

$$Z_{fjs}(t) = \alpha\left[1 - (1+\beta t)\cdot\exp\left\{-\beta t - \sigma w(t) - \sum_{i=1}^{Y_t(\lambda_1)} \log V_i^1 - \sum_{i=1}^{Y_{t'}(\lambda_2)} \log V_i^2\right\}\right]. \tag{5.15}$$

Considering the time delay over $t_2$ $(t_2 \geq t_1)$, we can formulate the flexible jump diffusion process models as follows:

$$Z_{fje}(t) = \alpha\left[1 - \exp\left\{-\beta t - \sigma w(t) - \sum_{k=1}^{K}\sum_{i=1}^{Y_{t^k}(\lambda_k)} \log V_i^k\right\}\right], \tag{5.16}$$

$$Z_{fje}(t) = \alpha\left[1 - (1+\beta t)\exp\left\{-\beta t - \sigma w(t) - \sum_{k=1}^{K}\sum_{i=1}^{Y_{t^k}(\lambda_k)} \log V_i^k\right\}\right], \tag{5.17}$$

where $t^k$ $(k = 1, 2, \ldots, K)$ means $k$ th specific time for major version upgrade, and $K$ are the number of major version upgrade.

## 5.3 Parameter estimation for noise parameters of flexible effort prediction models

### 5.3.1 Method of maximum likelihood for the drift term

We estimate several unknown parameters $\alpha$, $\beta$, $b$, and $\sigma_1$ in eqs. (5.16) and (5.17) in this section. Note that $\sigma_2$ and $l$ are the known parameters, because $\sigma_2$ and $l$ are obtained as the network factor. The joint probability distribution function for $Z(t)$ is defined as

$$P(t_1, y_1; t_2, y_2; \ldots; t_K, y_K) \equiv \Pr[Z(t_1) \leq y_1, \ \ldots, \ Z(t_K) \leq y_K | Z(t_0) = 0]. \tag{5.18}$$

From. eq. (5.18), the probability density is shown as follows:

$$p(t_1, y_1; t_2, y_2; \ldots; t_K, y_K) \equiv \frac{\partial^K P(t_1, y_1; t_2, y_2; \ldots; t_K, y_K)}{\partial y_1 \partial y_2 \cdots \partial y_K}. \tag{5.19}$$

Then, the likelihood function, $\lambda$, for the actual effort data $(t_k, y_k)(k = 1, 2, \ldots, K)$ is constructed as

$$\lambda = p(t_1, y_1; t_2, y_2; \ldots; t_K, y_K). \tag{5.20}$$

As the technique for mathematical manipulations, we consider the following logarithmic likelihood function:

$$L = \log\lambda. \tag{5.21}$$

Then, the estimates $\alpha^*$, $\beta^*$, $b^*$, and $\sigma_1^*$ are estimated by using the following simultaneous likelihood equations. Then, the estimates are given by making $L$ in eq. (5.21) maximize

$$\frac{\partial L}{\partial \alpha} = \frac{\partial L}{\partial \beta} = \frac{\partial L}{\partial b} = \frac{\partial L}{\partial \sigma_1} = 0. \tag{5.22}$$

## 5.3.2 Genetic algorithm approach for the jump terms

Our model includes the mixed distribution such as jump diffusion and Wiener processes. Therefore, the complicated likelihood function is structured from our model. Then, it is difficult to estimate the unknown parameters of jump terms of our model. Also, several estimation methods of unknown parameter included on jump terms have been proposed in the past. However, the novel estimation method has only a few presented. We discuss that the estimation methods based on GA in order to estimate the unknown parameters in the jump terms in this section. The estimation procedure of unknown parameters of our jump model is as follows:

Then, we consider $\gamma$, $\mu$, and $\tau$ as the unknown parameters in the jump term. Then, the parameters $\mu$ and $\tau$ structure the jump range $V_i$.

In particular, the fitness function is structured from the estimate and actual data. The following error function as the fitness function is defined in this chapter, for example, the error between the estimate and the actual values:

$$\min_{\theta} F_i(\theta),$$
$$F_i = \sum_{i=0}^{K} \{Z_j(i) - y_i\}^2,$$

where $M_j(i)$ is the cumulative software operation effort at time $i$ in our model based on jump diffusion process, $y_i$ the cumulative software effort. Also, θ means the parameter set in terms of $\gamma$, $\mu$, and $\tau$.

The above-mentioned processes are applied to the unknown parameter of jump terms [11, 12].

### 5.3.3 Deep learning approach for the jump terms

We use the method of maximum likelihood for the unknown parameters $\alpha$, $\beta$, $b$, and $\sigma$. In particular, it is very difficult to make a decision of the unknown parameters of jump terms in our models because of the complexity in likelihood function including multiple distributions based on the Wiener process and jump diffusion one. Several estimation methods for jump parameters of jump diffusion process model have been proposed by the specified researchers. However, there are no effective methods of such estimation. This chapter discusses the estimation method of parameters in terms of jump terms. Then, a deep learning is used in this chapter in order to make a decision the jump parameters of the discussed model.

For example, we assume that our jump diffusion process models includes the parameters $\lambda_1$ and $\lambda_2$ for $Y_t$ and $Y_{t'}$, similarly, $\mu_1$, $\mu_2$, $\tau_1$, and $\tau_2$ for $V_i^1$ and $V_i^2$ in eqs.(5.16) and (5.17). Then, the set parameters $\boldsymbol{J}$ in terms of $\lambda_1$, $\mu_1$, and $\tau_1$ are estimated by deep learning algorithm in case of $(t \geq 0)$. Similarly, the set parameters $\boldsymbol{J}'$ in terms of $\lambda_2$, $\mu_2$, and $\tau_2$ make a decision by using the deep learning algorithm in case of $(t' \geq t_1)$.

The deep learning structure in this chapter is represented in Figure 5.1. In Figure 5.1, $z_l(l = 1, 2, \ldots, L)$, and $z_m(m = 1, 2, \ldots, M)$ means the pretraining units in hidden layer. Also, $o_n(n = 1, 2, \ldots, N)$ is the unit of output layer as the compressed characteristic. In this chapter, we apply the deep feedforward neural network from among several algorithms of deep learning in order to learn the OSS fault big data on bug tracking systems. We apply the following input data sets in terms of the each unit on the input layer. Then, the unknown parameters as the objective variable are given as the parameter set $\boldsymbol{J}$ in terms of $\lambda_1$, $\mu_1$, and $\tau_1$. The following nine items as explanatory variables are set to the units of input layer:

- Date and time
- OSS product name
- OSS component name
- Name of OSS version
- Reporter (nickname)
- Assignee (nickname)
- Fault status
- OS name
- Level of severity in terms of fault

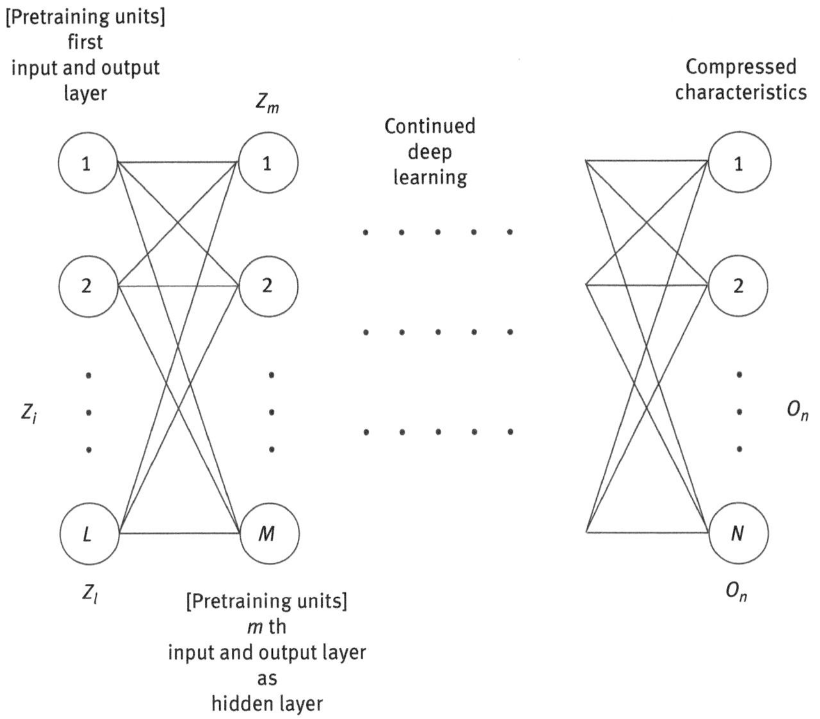


**Figure 5.1:** The feed forward deep neural network structure as deep learning.

Then, the input value for each input units will be changed from the character to the numerical value [13].

## 5.4 Numerical examples

The Apache HTTP Server [14] from Apache Software Foundation well-known as OSS. Figure 5.2 is the estimated operation effort expenditures based on exponential effort prediction model by using the GA. Also, the dot–dash line shows the start line of beta version 7.0.0 major-version-upgraded from version 6.x line. From Figure 5.2, we find that the jump becomes large after the time 1,826 days. Similarly, Figure 5.3 represents the estimated cumulative OSS operation effort expenditures based on S-shaped effort prediction model by using the GA. From Figure 5.3, we find that the S-shaped effort prediction model fits better than exponential effort prediction model for the actual data sets, where the long-dash line shows the start line of version 4.1.31 major-version-upgraded from version 3.x line in In Figures 5.2 and 5.3.

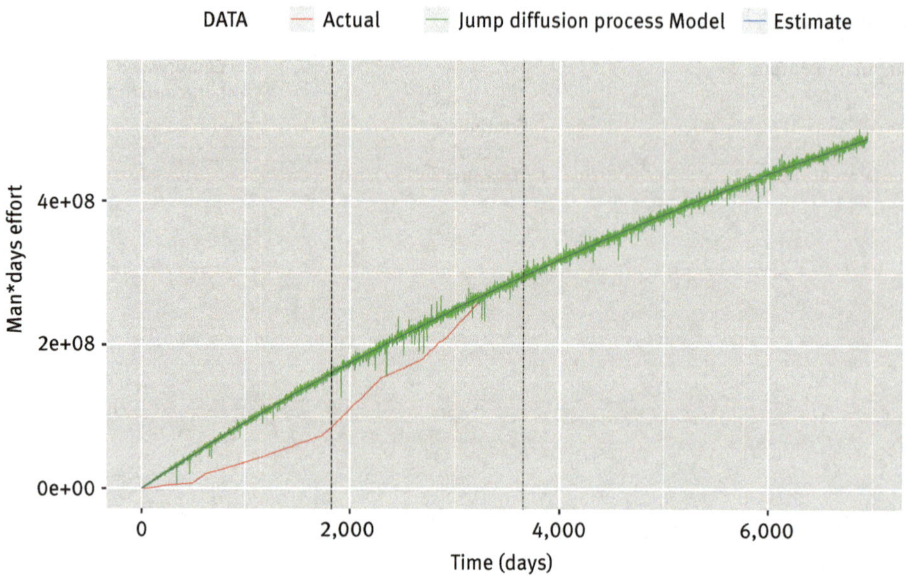


**Figure 5.2:** The estimated cumulative OSS operation effort expenditures based on exponential effort prediction model by using the GA.

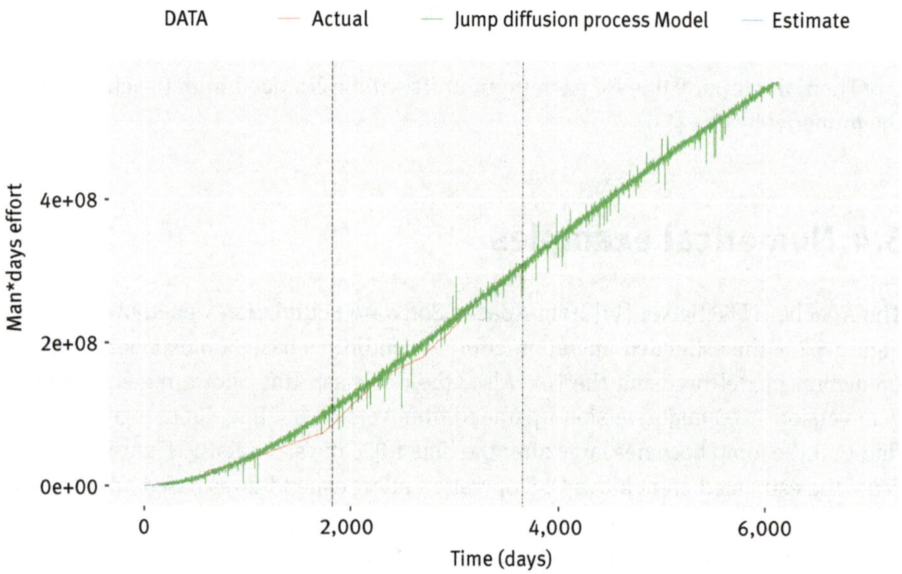


**Figure 5.3:** The estimated cumulative OSS operation effort expenditures based on S-shaped effort prediction model by using the GA.

Similarly, Figures 5.4 and 5.5 show the estimated cumulative OSS operation effort expenditures based on exponential effort prediction model by using the deep learning, and the estimated cumulative maintenance effort expenditures based on S-shaped effort prediction model by using the deep learning. In particular, the data sets of specified phases are estimated by the deep learning in Figures 5.4 and 5.5. From Figures 5.4 and 5.5, we found that the estimates by deep learning can show for each phase in detail.

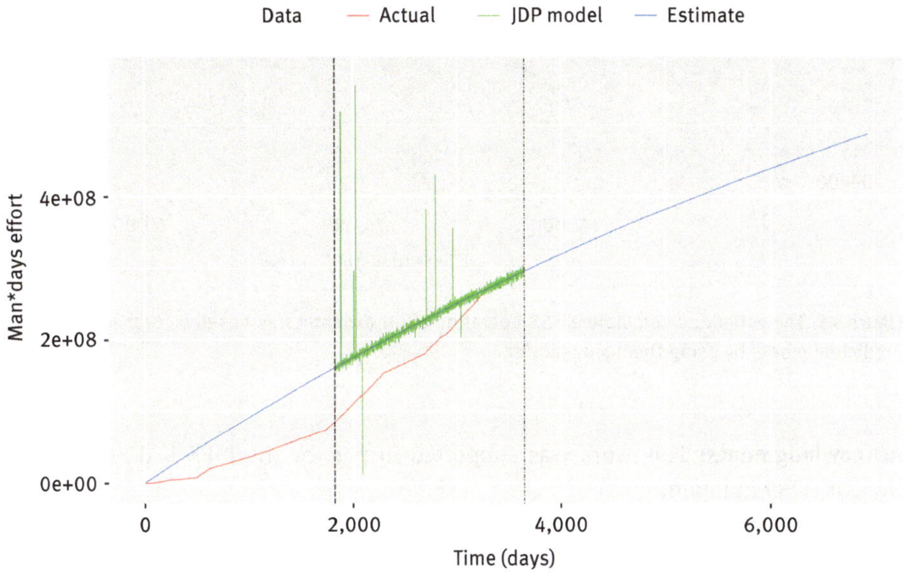


**Figure 5.4:** The estimated cumulative OSS operation effort expenditures based on exponential effort prediction model by using the deep learning.

## 5.5 Concluding remarks

This chapter focusses on the software effort control for OSS projects. The optimal estimation of OSS effort indirectly relates to the OSS quality, OSS reliability, and OSS cost reduction. In this chapter, we have discussed the method of OSS effort assessment considering the irregular situations with jump term from the characteristics including several OSS version upgrade. It is difficult for the OSS project managers to estimate the many parameters of jump terms. Then, we discussed several methods of parameter estimation in our effort prediction models. The proposed parameter estimation methods will be useful as the estimation method of the progress with OSS version upgrade from the standpoint of the project management.

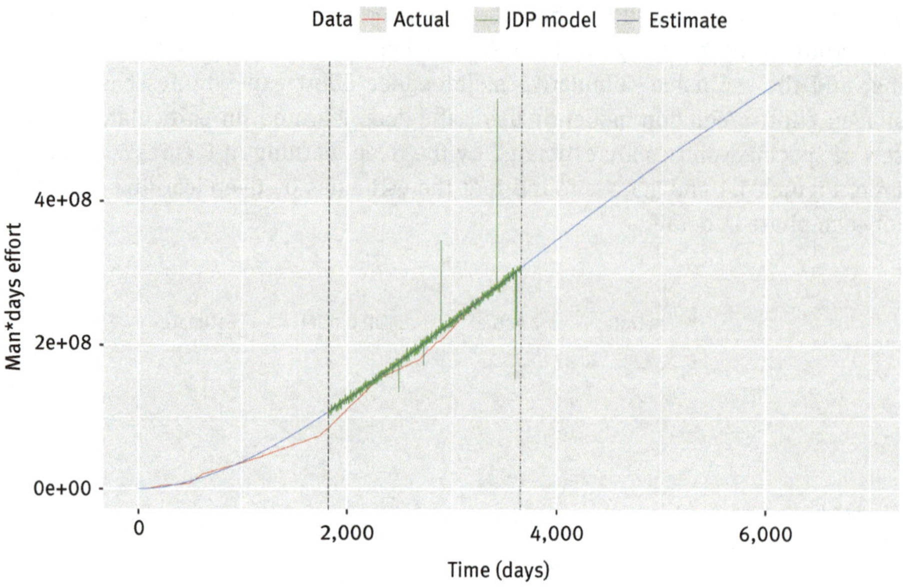


**Figure 5.5:** The estimated cumulative OSS operation effort expenditures based on S-shaped effort prediction model by using the deep learning.

**Acknowledgments:** This work was supported in part by the JSPS KAKENHI Grant No. 20K11799 in Japan.

## References

[1] Yamada, S. and Tamura, Y. (2016). OSS Reliability Measurement and Assessment, Springer International Publishing, Switzerland.

[2] Norris, J. (2004). Mission-critical development with open source software, IEEE Software Magazine, 21(1), 42–49.

[3] Zhou, Y. and Davis, J., "OSS reliability model: an empirical approach," Proceedings of the Fifth Workshop on OSS Engineering, pp. 67–72, 2005.

[4] Yamada, S. (2014). Software Reliability Modeling: Fundamentals and Applications, Springer-Verlag, Tokyo/Heidelberg.

[5] Lyu., M.R. ed. (1996). Handbook of Software Reliability Engineering, IEEE Computer Society Press, Los Alamitos, CA, U.S.A.

[6] Musa, J.D., Iannino, A. and Okumoto, K. (1987). Software Reliability: Measurement, Prediction, Application, McGraw-Hill, New York.

[7] Kapur, P.K., Pham, H., Gupta, A. and Jha, P.C. (2011). Software Reliability Assessment with OR Applications, Springer-Verlag, London.

[8] Arnold, L. (1974). Stochastic Differential Equations–Theory and Applications, John Wiley & Sons, New York.

[9] Wong, E. (1971). Stochastic Processes in Information and Systems, McGraw-Hill, New York.
[10] Yamada, S., Kimura, M., Tanaka, H. and Osaki, S. (1994). Software reliability measurement and assessment with stochastic differential equations, IEICE Transsactions on Fundamentals, E77–A(1), 109–116.
[11] Tamura, Y., Sone, H. and Yamada, S. (2019). Productivity assessment based on jump diffusion model considering the effort management for OSS project, International Journal of Reliability, Quality and Safety Engineering, 26(5), 1950022-1–1950022-22. World Scientific.
[12] Tamura, Y. and Yamada, S. (2019). Maintenance effort management based on double jump diffusion model for OSS project, Annals of Operations Research, Springer, US, Online First, 1–16. Doi: 10.1007/s10479-019-03170-w.
[13] Tamura, Y. and Yamada, S. (2017). Fault identification and reliability assessment tool based on deep learning for fault big data, Journal of Software Networking, 2017(Issue 1), 161–176. Doi: 10.13052/jsn2445-9739.2017.008.
[14] The Apache Software Foundation, The Apache HTTP Server Project, http://httpd.apache.org/

Mario Diván and María Laura Sánchez-Reynoso

# 6 Modeling the meaning of data streams and its impact on the system performance

## A perspective of the data stream content in the data-driven decision-making

**Abstract:** Under the category of the data stream engine, it is possible to find software that is able to process data arriving in real-time from different data sources. In this sense, the Internet-Of-Things (IoT) has played a key role because it provides the bases for implementing a cheap and interesting alternative for data collecting, deploying different kind of devices along the monitoring field. However, IoT is characterized for the heterogeneity, which is associated with different kinds of devices, proprietary data formats related to each device, jointly with the differences between precision and accuracy depending on the associated technology. This level of heterogeneity is useful because it allows implementing nonproprietary solutions and be able to choose between wider alternatives, but also incorporates an additional level of complexity derived from its heterogeneity. As the main contribution, the idea of cooperative and exclusive data streams jointly with its impact on the system performance is explained. Thus, an alternative to model the structure and the meaning related to data streams related to measurement coming from IoT devices are outlined. On the one hand, the modeling strategy pretends to describe the heterogeneous data sources through which a set of measures are jointly informed under the same data stream. On the other hand, the modeling strategy in those cases in which each metric is individually informed is described. Both perspectives are mutually exclusive because the first one interprets each stream as a common channel through which different metrics and measures are informed (i.e., a cooperative data stream), while in the second one, each data stream is exclusive for a given metric and their associated measures (i.e., an exclusive data stream). An alternative for modeling each one and to translate the definition between them is presented. In addition, a simulation associated with the overhead associated with both kinds of data processing is described.

**Keywords:** data streams, data meaning, real-time data processing, system performance, exclusive data streams, cooperative data streams

**Mario Diván, María Laura Sánchez-Reynoso,** National University of La Pampa, La Pampa, Argentina

https://doi.org/10.1515/9783110619058-006

## 6.1 Introduction

The dynamism and adaptability are two properties of the system introduced in the general system theory [1], which enable it to be adjustable according to the contextual changes. On the one hand, the adaptability refers to the system's ability to adequate to its environment in order to follow satisficing its aim. On the other hand, the dynamism is associated with the velocity in which a system is able to adapt to a contextual change. Both properties play an essential role in relation to system performance and the ability to manage the available resources [2].

Globalization tends to establish interdependence among countries from the economical point of view, which implies (among other things) the necessity to be constantly communicated between stakeholders to keep the link in an appropriated way [3]. In this sense, communication technologies played an essential role, and today, to be communicated with people around the world is quite simple and accessible for any citizen.

In this sense, the decision-making processes have changed for adapting to a globalized scenario, extending its boundaries and turning on a distributed context. Extending the boundaries enable to incorporate new perspectives and point of view related to cultural and social factors, at the same time in which the distributed context provides risk and responsibility distribution among the players, avoiding the concentration and fostering the balanced and wide participation.

A globalized environment requires a sensible orchestration of information that needs to be available to each stakeholder in order to support the data-driven decision-making. Intuition is important in some situations in which a person needs to react to deal with the uncertainty. However, when it is possible to have enough information for reducing the uncertainty, it always will be a better option because each decision is based on previous knowledge or experiences. The orchestration implies a certain level of synchronization in order to warranty that each part has enough and updated data for deciding.

The real-time decision-making elevates the level of required synchronization up to an extreme, implying that each decision should be sustained by the last known data coming in a direct way from the source. It sounds easy to say but it has a lot of challenges related to (i) Data collecting: the way in which each data is obtained; (ii) Data uality: it is related to different data source aspects such as the confidence, accuracy, and precision; (iii) Data transporting: it refers to the way in which the data are carried from the data source to each stakeholder involved throughout the decision-making; (iv) Data processing: It indicates the way in which each data is processed in order to support the decision-making, keeping in mind that new data are continuously arriving and the processing resources are limited (i.e., memory, processor, etc.); and (v) Decision-making process: it focuses on the used decision-making schemas in a distributed environment [4, 5].

Data stream incorporates a continuous data processing paradigm able to deal with heterogeneous data sources providing data such as they are. The data sources are autonomous; they can generate an unbounded data stream varying the arriving rate without any previous notice. The paradigm supposes that a data stream engine is able to take different data sources and to process them in different ways (e.g., synthesizing, joining, grouping, etc.) with the aim of generating new outcomes or even answer different kinds of queries (be it planned or not). However, an important difference with traditional data management is that in data streaming the data are processed at the arriving time, being discarded after that. That is to say, the data is useful and valuable when they arrive because they are describing the situation in a given instant, however, always there will be newly updated data coming for being processed and analyzed in order to describe the evolution of the situation being monitored [6, 7].

Internet-of-Things (IoT) is a concept related to heterogeneous, tiny, available, and cheap devices able to be used as data collectors, allowing a wide application area. This versatility derives in a challenge given by the complexity incorporated through the heterogeneity itself [8]. That is to say, different devices will have different data formats, precision, accuracy, reliability, among other aspects that coexist under the same collecting strategy articulated through network communications. In addition to the heterogeneity and considering the wide application of this kind of devices, challenges related to security [9], articulation with Fog computing [10], Big Data [11], Blockchain [12], among others. The versatility and accessibility of IoT devices as data collectors added to the real-time processing abilities related to data streaming, foster an ideal scenario for data analytics or monitoring applications in general [13].

As main contributions, (i) Exclusive and cooperative data streams are introduced as a way for modeling the heterogeneity from the data sources, allowing discriminating from the beginning the kind of expected behavior in relation to each data source jointly with the associated processing requirements; (ii) A translating schema between exclusive and cooperative data streams is shown with the aim of making both interoperable; and (iii) An analysis of the potential overhead associated with the translating schema between cooperative and exclusive data streams is outlined.

The chapter is structured in nine sections. Section 6.2 describes some approaches related to the data organization associated with the data steaming's environment. Section 6.3 introduces the systematic mapping of literature on data stream modeling. Section 6.4 describes the necessity of a framework throughout the measurement process. Section 6.5 synthesizes the idea related to the processing strategy. Section 6.6 introduces the modeling of the exclusive and cooperative data streams respectively. Section 6.7 outlines some basic operations over exclusive and cooperative data sources, while Section 6.8 analyzes the processing overhead associated with it. Finally, some conclusions and future works are presented.

## 6.2 Current approaches of the data organization in streaming

In [14] the authors propose an architecture for recreating data streams from web data and/or sensors' data stored on the Cloud. They indicate that the underlying idea is to take advantage of the huge volume of data available on the web, but they cannot be directly queried because it does not contain an associated API. Thus, the architecture named Sensorizer is able to get heterogeneous data from the web by means of containers that are able to send under an integrated stream all the data together. The concept of virtual sensor nodes is introduced with the aim of implementing multiple virtual transducers, where each transducer is related to single web content. Thus, each virtual sensor node could be associated with a lot of heterogeneous data sources that will provide data according to the monitored web content.

In [15] authors synthesize Facebook's ecosystem associated with real-time data processing. In that context, data coming from users and their activities are processed and analyzed at the light of performance, fault tolerance, scalability, correctness, and ease of use. In this case, all the processing schemas are related to heterogeneous data coming from different systems in an integrated way, for that reason, the original data stream and those generated from them could be related to different user data sources.

Up to here, the analyzed data streams in the previous proposals correspond to heterogeneous data that are informed through a common media that is understood as a data stream. On the one hand, in [14] the data streams are integrated from data consumed from the web depending on the containers and defined transducers. On the other hand, in [15] the streams are consumed, processed, and derived from users' activity jointly with involved system information. In [16] a load-aware shedding algorithm applied to data stream systems is proposed. However, the data stream is understood as an unbounded sequence of "tuples," where the tuple is a set of (key, value) pairs that could be combined for supporting complex data structures. That is to say, this case incorporates a certain level of structuration when the previous ones did not establish explicit restrictions around it. This is particularly important to mention because it does not seem to exist in homogeneity around the concept of the data stream, however, there is a consensus around the idea of data streams that could be synthetically defined as an unbounded data sequence.

In [17] the data stream is referred to as an unbounded sequence of real-time data, where the data is understood as a tuple. Each tuple has attributes characterizing some aspect jointly with a special attribute that represents the timestamp. In other words, the data stream is supposed ordered in terms of its timestamp. This perspective of tuple implies a bidimensional point of view where the attributes play the role of columns, while each transmitted record with that structure represents a tuple or record itself. This perspective of the data stream has a certain level of structuration in terms of the relational paradigm, generating different questions, for example, when the data is a likelihood distribution related to one attribute, is the sequence of pairs (value, likelihood) informed into one tuple or by means of a set of tuples? In the first case, what data model is employed for interpreting the data organization? In the last case, how can the relationship be established for indicating that all the tuples correspond to the same likelihood distribution? In some situations, the business logic is embedded into the streaming application, which implies a dependence on the written code in place of the data meaning, mixing the data layer with the processing layer. The last represents a risk because of the coupling between layers is increased in place of fostering the uncoupling, which would make easy the applications' maintenance.

In [18], a model for the online reconstruction of sessions is introduced, articulating the batch environment jointly with the data stream context. Data from the logs are incorporated through the figure of records, which represents a similar meaning to the tuple. However, in this case, authors do not restrict the record to a bidimensional data organization, being possible to integrate a record with different elements coming from the logs.

In [19], a survey related to GeoStreams is introduced describing it such as a data stream containing temporal and spatial data. However, the data stream is defined as a permanently updating source of data coming from active origins. An interesting difference in comparison to the previous perspectives is the term "active origins," which explicitly highlights the push mechanism related to the data generators, and the independence between the data generated in the source with respect to the data once they enter the data stream engine. As a corollary, the data coming from the active origins are not plain, that is to say, it is not just a number. On the contrary, it is complex data that contains temporal and spatial information that is useful in these kinds of geographic applications.

In [20] the Spark Structured Streaming is introduced. The new model simulates a live data streaming such as a bidimensional table that always is growing up. Thus, the concept of a tuple is associated with a set of attributes or columns, while the table is composed of a set of tuples. The particularity of this new concept of table is that it has no end, always there will be a new tuple or record to append. This incorporation is a complement of the previous streaming architecture oriented to provide user facilities and to make easy the possibilities of application.

In [21] a session reconstruction schema is outlined based on the servers' logs. The underlying idea is to collect data from different logs, integrating and processing them with the aim of recreating user sessions for being used in mining process models. In this case, even when the native source is active, the data generation is not primarily oriented to low latency schema. For that reason, a continuous reading, processing, and adapting from the logs need to be addressed for generating the data stream in a continuous way. Also, the data streams are not necessarily integrated by a single value, but also it could be composed of semistructured content.

In [22] a library named IoTPy with the aim of helping to develop stream applications in an easier way is presented. In this context, the data stream is defined as an unbounded sequence of items, in where the stream itself cannot be modified in any way. The only enabled operation on the data stream is the incorporation of a new item at the end of the data stream. It initially outlines a difference with the previous proposals in two perspectives: (i) The first one is that the proposal refers to items and not tuples. The item is a value collected from an IoT sensor, which implies that the data stream is an unbounded sequence of values coming from the sensors, and (ii) The second one indicates that each data stream is related to data coming from a specific sensor, which implies that only data from that sensor will be enqueued and no other.

In [23] a hash table data structure for detecting duplicate on data streams is introduced. This perspective is interesting to be considered because it needs to understand the data organization of each data stream's element to indicate whether it is duplicated or not. Thus, the authors define the data stream as an infinite stream of symbols, which introduces a meaning closer to the Chandy proposal [22]. That is to say, Géraud talks about symbols while Chandy refers to items. Both seem to be in contraposition to the tuple concept or semistructured content previously presented in other proposals.

## 6.3 Data stream modeling: current trends

With the aim of identifying the main tendencies in relation to data stream modeling, a Systematic Mapping Study (SMS) was carried out based on "Data Stream Modelling" as the main terms. The steps and procedures for implementing SMS are described in [24, 25].

Briefly, the SMS's basic steps are organized around six stages. The first stage establishes the aim of the study. The second stage describes some research questions aligned to the defined objective. The third step is to define the search strategy, indicating the terms to be queried. The fourth stage describes the data extraction stage through which one gets descriptive data about the listed articles into the result, obtaining characteristics that are useful for filtering. In the fifth phase, synthesis is carried forward around the extracted data. Finally, the assurance of results is monitored.

In this case, the aim is established to identify different modeling alternatives of the data stream from the structural point of view. That is to say, our idea is to analyze tendencies related to how the structure of the data flow is modeled in a context associated with heterogeneous data sources, which need to be processed in real-time.

The Research Questions (RQ) derived from the indicated objective could be synthesized as follows: (1) RQ1: What kind of data organization the data stream has?

(2) RQ2: How the processing is affected by the data organization? On the one hand, the motivation related to the first research question focuses on the way in which the structure (i.e., the data organization) allows jointly integrate or not data coming from heterogeneous data sources under the same stream. On the other hand, the second research question is oriented to analyze the impact that one organization or others have on the data processing itself.

In this way, defined the objective jointly with its associated research questions, it is possible to write the basic search string. Thus, the search string is indicated as "data stream modelling," and it was performed on the Scopus database. The considered documents were articles, book chapters, and conference papers between 2016 and 2020, written in English and corresponding to the computer science field. This is important because the underlying idea consisted in analyzing the current situations and the early publications related to the next year in terms of trends. Table 6.1 synthesizes the search string performed on the Scopus database.

**Table 6.1:** Search string performed on the Scopus database on November 15 of 2019, 20:05 (Argentina time zone).

| Main term | Alternative terms | Performed search query on Scopus |
|---|---|---|
| "Data Stream Modelling" | "Real-Time Data Modelling" or "Data Flow Modelling" | TITLE-ABS-KEY ( "Data Stream Modeling" OR "Real-Time Data Modeling" OR "Data Flow Modeling") AND ( LIMIT-TO ( DOCTYPE, "cp") OR LIMIT-TO ( DOCTYPE, "ar") OR LIMIT-TO ( DOCTYPE, "ch")) AND ( LIMIT-TO ( SUBJAREA, "COMP")) AND ( LIMIT-TO ( LANGUAGE, "English")) AND ( LIMIT-TO ( PUBYEAR, 2020) OR LIMIT-TO ( PUBYEAR, 2019) OR LIMIT-TO ( PUBYEAR, 2018) OR LIMIT-TO ( PUBYEAR, 2017) OR LIMIT-TO ( PUBYEAR, 2016)) |

Thus, 30 documents were obtained, and their metadata were exported to an Excel file for its analysis (e.g., DOI, authors, publishing, abstracts, and keywords). From there, an individual reading was made from the abstract to determine whether each one was applicable to the specific subject or not. This is important, because the initial searching on the database is syntactical, that is to say, the search engine looks for the specific indicated terms, but it does not analyze the meaning of each word in the context. For this reason, the individual reading of the abstract or full text is a key stage.

Once the main filters (e.g., the period) were applied (see Table 6.1), the filtering stage is associated with specifying the retaining and exclusion criteria. In this case, the inclusion criteria refer to those works that explicitly describe aspects of the data organization in the data stream or real-time data processing. However, keynotes, invited talks, and surveys are not considered, they will be rejected. Thus, from the original 30 papers in the answer, only 10 were retained in which their metadata are synthesized in Table 6.2 introduced in the next page.

Figure 6.1 synthesizes the evolution related to the kinds of publications for this specific subject, highlighting the associated publishers. It is important to observe that only three from the ten records correspond to conference papers, being the highest proportion associated with journals. That suggests that concerning this subject, the publications have a certain level of maturity in terms of its associated results.

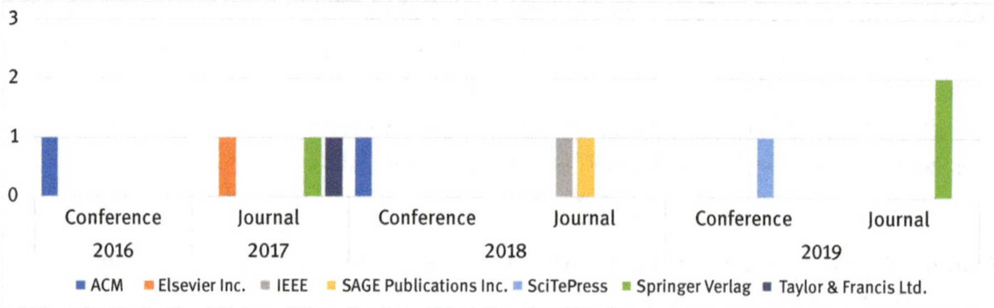


**Figure 6.1:** Evolution of the kinds of publication by year highlighting their associated publishers.

The idea is to analyze how each of the earlier-listed papers deals with the data stream concept from the morphological point of view and its impact on the data stream processing. Identifying the main tendencies about data stream modeling between 2016 and 2020 is especially interesting, because it is possible to appreciate the last perspectives and alternatives to model the data stream itself, independently of the previous interpretations.

Lughofer et al. [26] introduce a proposal of architecture oriented to the generalized evolving fuzzy systems. The underlying idea resides in an agile detection of data concept drifts without using additional parameters (e.g., a threshold). Data are received as a series of multidimensional vectors that are immediately analyzed, incorporating a rule splitting methodology working around an incremental way, which implies that each vector is read up to once time.

Lughofer [27] introduced advances, challenges, and tendencies of the online active learning and its mutual impact on between machine learning techniques and its context. Perspectives related to the classical batch off-line learning is contrasted to online active learning, exposing similarities and differences between them. In this context, data streams are understood as a sequence of data that could be sampled in different ways, employing different kinds of techniques. Different learning techniques are outlined and compared to them.

Georis-Creuseveau et al. [28] proposed a framework oriented to work around spatial data infrastructures. In this study, the data collecting strategy is based on the capturing of online questionnaires and semistructured interviews.

**Table 6.2:** Retained papers once the individual reading was made, considering the citation index on November 15 of 2019, 20:05 (Argentina time zone).

| Authors | Title | Year | Source | Publisher | First author's country | Cited by |
|---|---|---|---|---|---|---|
| Lughofer E., Pratama M., Skrjanc I. | "Incremental rule splitting in generalized evolving fuzzy systems for autonomous drift compensation" | 2018 | *IEEE Transactions on Fuzzy Systems* | IEEE | Austria | 23 |
| Lughofer E. | "On-line active learning: A new paradigm to improve practical usability of data stream modeling methods" | 2017 | *Information Sciences* | Elsevier | Austria | 18 |
| Georis-Creuseveau J., Claramunt C., Gourmelon F. | "A modelling framework for the study of Spatial Data Infrastructures applied to coastal management and planning" | 2017 | *International Journal of Geographical Information Science* | Taylor & Francis Ltd. | France | 8 |
| Kamburugamuve S., Wickramasinghe P., Ekanayake S., Fox G. C. | "Anatomy of machine learning algorithm implementations in MPI, Spark, and Flink" | 2018 | *International Journal of High Performance Computing Applications* | SAGE Publications Inc. | United States | 5 |
| Koek P., Geuns S. J., Hausmans J. P. H. M., Corporaal H., Bekooij M. J. G. | "CSDFa: A model for exploiting the trade-off between data and pipeline parallelism" | 2016 | *Proceedings of the 19th International Workshop on Software and Compilers for Embedded Systems, SCOPES 2016* | ACM | Netherlands | 1 |
| Dubrulle P., Gaston C., Kosmatov N., Lapitre A., Louise S. | "A data flow model with frequency arithmetic" | 2019 | *Lecture Notes in Computer Science (including subseries Lecture Notes in Artificial Intelligence and Lecture Notes in Bioinformatics)* | Springer Verlag | France | 0 |

(continued)

**Table 6.2** (continued)

| Authors | Title | Year | Source | Publisher | First author's country | Cited by |
|---|---|---|---|---|---|---|
| Meinig M., Tröger P., Meinel C. | "Finding classification zone violations with anonymized message flow analysis" | 2019 | *ICISSP 2019 – Proceedings of the 5th International Conference on Information Systems Security and Privacy* | SciTePress | Germany | 0 |
| Masulli F., Rovetta S. | "The Challenges of Big Data and the Contribution of Fuzzy Logic" | 2019 | *Lecture Notes in Computer Science (including subseries Lecture Notes in Artificial Intelligence and Lecture Notes in Bioinformatics)* | Springer Verlag | Italy | 0 |
| Chadli N., Kabbaj M. I., Bakkoury Z. | "Detection of dataflow anomalies in business process an overview of modeling approaches" | 2018 | *ACM International Conference Proceeding Series* | ACM | Morocco | 0 |
| Mackie I. | "A geometry of interaction machine for gödel's system t" | 2017 | *Lecture Notes in Computer Science (including subseries Lecture Notes in Artificial Intelligence and Lecture Notes in Bioinformatics)* | Springer Verlag | United Kingdom | 0 |

Kamburugamuve et al. [29] described the perspectives related to data management and its impact on platforms such as Apache Spark, Flink, and MPI. In general, they indicate that the platforms consider the data and their processing through different variations of graph model, while that data itself is conceptualized using a bidimensional and immutable data structures known as Resilient Distributed Datasets (RDD) in Spark or DataSet in Flink. The data stream could be varied throughout the processing line implemented in each engine, considering them as a sequence bidimensional of tuples that are input for a set of operation, while it could be the output for another set of operations.

Koek et al. [30] analyzed the challenge related to the data communication under a context of parallelism. Basically, the data streams are understood as a sequence of atomic data (e.g., an integer) able to communicate two or more tasks (i.e., operators). They propose a model for coordinating the data flow with eventual auto-concurrency between operations throughout the whole data processing.

Dubrulle et al. [31] outlined a proposal for modeling data pipelines with restrictions between producers and consumers. The graph theory is used for representing as edge a consumer or producer, that it will be determined based on the sense of the arc. When the arc arrives at an edge, the edge will be consuming the data, but when the arc is departing from an edge, it will play a producer role. Annotations could be employed for discriminating the behavior between different data sources (e.g., the data frequency). Also, the challenge related to the data fusion unifying the data produced from a set of heterogeneous data sources is introduced.

Meinig et al. [32] dealt with the challenge of identifying the classification zone violation. For that purpose, the logs from servers are used as input. However, given the heterogeneity and different data formats related to each one, a data conversion stage is necessary. In this sense, the authors describe basically the necessary operations for translating the unstructured raw data in each log to a structured data organization able to be used for generating models. Thus, the raw data are converted into a structured and bidimensional sequence of data following the idea of tuples characterized through a set of attributes.

Masulli et al. [33] described the challenges associated with the big data environment, its projection over time, and the role of the fuzzy logic. They differentiate two dimensions: data itself and the content. On the one hand, the data refer to the captured fact, that is to say, those records that need to be kept in an organization for eventual using. On the other hand, the content talks about the data meaning, its role, and impacts on the knowledge. The authors make focus on the necessity of developing tools oriented to data stream modeling, an aspect that has taken a significant role in past times. For additional details, interesting perspectives from authors can be analyzed in terms of the clustering of nonstationary streams [34] and tracking time-evolving data streams [35].

Chadli et al. [36] proposed different approaches for dealing with the data flow associated with the business processes jointly with their associated challenges around the anomaly detection (e.g., missing values). The work refers to data items embedded in the data flow, which implies a sequence of tuples organized as a set of columns flowing between processes. They analyze the using of a data-flow matrix for studying the challenge, while the Petri's net is used for anomaly detection throughout the dataflow exchanged between processes.

Mackie et al. [37] proposed a dataflow model. They talk about tokens for representing the data that is traveling throughout a network. The underlying idea of the sequence is associated with an atomic sequence of data being communicated between computation components that are previously determined.

## 6.4 Measurement: the necessity of a framework

It is possible to define the measurement as the process in which an object or subject under analysis needs to be quantified through one or more attributes that help to characterize it. The measurement consists of a quantification schema where the idea is to contrast the obtained value about some known pattern. For example, the height of a person requires having a reference pattern such as the meter, the weight of a car requires to know the idea of the kilogram, and so on. These examples outline two different challenges. On the one hand, the necessity of quantifying a certain number of characteristics related to an object, subject, or concept. On the other hand, the necessity of having a reference pattern useful for comparison.

These challenges carry us to ask questions such as, Why we need to measure? and Why we need to compare the result with some pattern? Instinctively, persons tend to analyze each behavior, phenomenon, object, concept, and/or subject that surround them. The evolution of the human being required a deep understanding of the whole context, the different kinds of interactions and comprehension of each role played over time. The necessity of quantifying is early discovered in our world as a key asset associated with the progress of different activities, for example, the idea of knowing what is cool and hot, cheap and expensive, high and low, and heavy and light.

One of the first concepts developed by humans was the number. That concept answered the necessity of accounting each object available in the context (e.g., food). That was how the quantification was early applied for knowing the volume of things available. However, the requirement of comparing previous values with current values emerged as a necessity. That is to say, being possible to represent a concept (e.g., animals) through a number indicating the volume of available things, now the necessity falls into knowing whether the current quantity is upper than before or not. Thus, the comparison between magnitudes was part of a natural process [38].

Thus, the answer to why we measure for is associated with quantifying a characteristic and from each quantified value, a comparison pattern is employed for determining whether a concept has been increased or not. The answer to why a comparison pattern is necessary, it falls into the requirement of establishing a common and uniform reference useful for anyone who needs to establish an analogy with a concept. It was one of the original motives for creating, for example, the metric system, among others.

Nowadays, the measurement as a process is applied in different fields, and it could be indicated as a transversal and nonexclusive practice. In effect, the measurement process could be used from computer science to life science on different concepts, from the processor's temperature to the heartbeat rate respectively. This aspect constitutes a huge virtue but at the same time a challenge, because the ability to be deployed on different fields implies the ability to deal with the heterogeneity, which impacts directly on the problem complexity. Imagine for a moment the different kinds of devices that could be used in different situations, the associated accuracy, required precision, methods, and so on. This allows introducing the context in which each measurement process is performed, that is to say, the employed instruments, methods, and concept under study is not isolated from the world, by the contrary, all of them they are embedded in their specific environment, mutually affecting each other.

Up to here, it is worthy to mention that the measurement is focused on a quantification that needs to be compared with a certain pattern, which is influenced through the environment and affected by the involved heterogeneity. However, something that seems to be easy not always is trivial to implement. For example, it is possible to try to measure the corporal temperature of a person using a thermometer, but one thing is to get a measure with a device positioned in the internal region of the ear, and another very different thing is to get the measure from the axillar region. Even with the same device under the same environmental conditions, the result could vary depending on the employed method [39]. This is important to highlight because the essence of the measurement is the comparability of values.

In this way, it is possible to appreciate that the intention of measuring to quantify a concept has a set of involved agreements. In other words, the concept related to the measurement needs to be identified jointly with the characteristics that describe it. From the descriptive characteristics, how each quantitative value is obtained should be defined. Also, the strategy for making comparable each value over time is a critical thing considering the necessity to determine levels of evolution or not based on the patterns taken as a reference.

The current world and associated economies require a measurement process as agile as possible, tending to be closer to the real-time data processing than before. The distributed, heterogeneous, dynamic, unpredictable, and unexplored environment in which the transactions currently take place imposes a set of additional conditions that need to be attended for ensuring the comparability [40]. For example, the fact of measuring stock variations based on different markets could be not directly comparable because each market could have associated particularities (e.g., volatility, regulations, etc.).

Due to the expected times for the measurement process, the idea is to approach an agile alternative as extensible, reliable, and stable as possible. Thus, the automatization of the measurement process is a must for governments, businesses, and organizations in general, which need to optimize different budgets in order to reach their respective aims.

There are a lot of concepts that need to be agreed before implementing a measurement process, for example, the underlying idea of metric, measure, measurement, scale, unit, method, among other concepts jointly with the relationship between them. This constitutes an important aspect because if two persons understand different things for the concept of metric, the value comparability would be directly affected.

For that reason, before implementing or automatizing any measurement process, the fact of sharing and agreeing with the involved concepts to be used throughout the process is critical for ensuring the whole reliability of the system [41]. Figure 6.2 synthesizes a global perspective of the involved concepts in the measurement process.

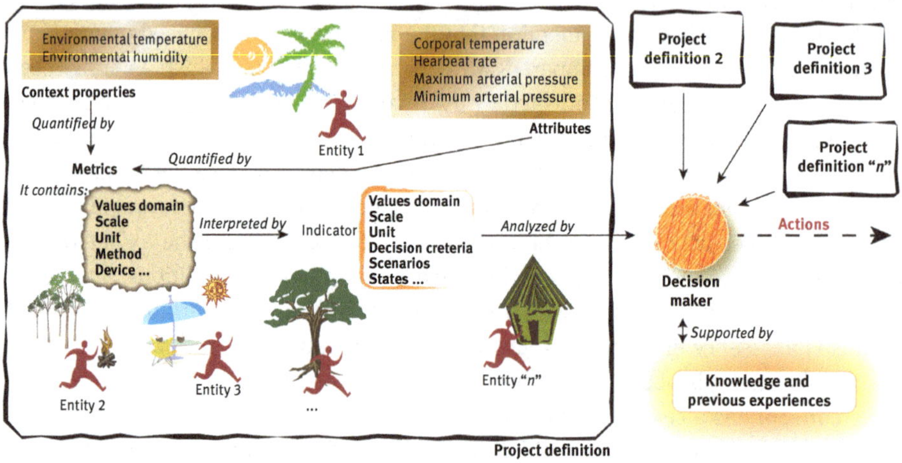


**Figure 6.2:** A Global perspective of the involved concepts in a measurement process.

As it is possible to appreciate from the previous figure, the concept to be monitored (entity in Figure 6.1) needs to be described by a set of attributes (e.g., heartbeat rate). This implies a discrete representation of the aspects to be analyzed for the given entity. In addition, the entity is immersed in a context, which is characterized using the context properties (e.g., environmental temperature). The last ones represent a discrete interpretation of the characteristics of the environment in which each monitored concept is acting. Both attributes and context properties are studied together to analyze the mutual incidence on each one.

Each attribute or context property is quantified by means of a metric. Each one has the expected values domain, associated scale, unit, method to be used to get the quantitative value, the device employed jointly with the method, and so on. The numerical value obtained from a metric is known as a measure. This definition is essential because it allows to know whether two measures are comparable or not. For example, the previous example related to measuring the corporal temperature would arrive at no comparable measures due to the metric uses different methods (i.e., axillar method versus in-ear method).

However, each measure only indicates a numerical value, but it does not say anything about how to interpret it. With that aim is incorporated into the concept of the indicator. The indicator consumes one or more measures, incorporating the necessary decision criteria based on the entity's states and current scenarios for analyzing the magnitude in context. Each interpretation will provide the interpretation but does not say much about the courses of action to follow.

The figure of the decision-maker reviews the interpretation of each indicator supported by the previous experiences and knowledge from experts, which is pertinent to the measurement project. Thus, it is able to provide some recommendations and/or courses of action to be implemented. For example, it could send an alarm to entity 1 indicating that the relationship between heartbeat and environmental temperature is near to a dangerous threshold.

As it is possible to appreciate, the previous experiences are shared throughout the different active project definitions, even being possible that each entity depending on its project definition, be different and it has different defined states and scenarios. All this heterogeneity and complexity needs to be defined under a consistent way, interpretable for humans and machines when the aim is the automatization of the measurement process.

This is just one interpretation of the involved concepts related to the measurement process, and it is highly possible that there exist other variations or definitions. Independent of that, the key asset is that each set of definitions is well-defined and organized to be understandable, communicable, and sharable for anyone who needs to use it. This point is where the framework is essential. The measurement framework must provide all the terms, concepts jointly with all the available relationships that are essential to implement a measurement process. For example, an alternative to formalize one is by means of ontology, making it communicable and processable.

Thus, once the framework is agreed, the measurement process could be effectively implemented in an automatic or manual way to ensure the repeatability, extensibility, and consistency of the process. The repeatability constitutes an essential aspect to know that new measures can be obtained following the same process definition. The extensibility refers to the possibility to add new requirements or update the previous ones keeping the descendant compatibility. The consistency implies that any change made on the project definition does not affect negatively the measures comparability.

## 6.5 Choosing the processing strategy

Choose the point throughout the processing chain in which the data starts to be modified, summarized, or transformed in any way has a huge impact on the whole data collecting strategy. That is to say, on thing is to try to process data on the same collecting device, while other thing is to process the data in a central virtual unit far from the data collectors. The first difference resides on the computing power necessary in each location (i.e., be it central or near to data sources), while the second aspect is the traveled path by the data. Farther the data are from the data source, bigger the data traffic and the resources consuming on the network are.

Each decision associated with the place where the processing could start has its advantages and disadvantages. On the one hand, when the data processing is near to the data sources, the data volume interchanged along the network is decreased but the processing requirements and storage capacities on mobile devices are increased. Also, the data processing needs to have a certain level of independence about other data sources, that is to say, the idea of processing data as near of sensors as possible implies a low level of dependence (i.e., coupling) on other sensors that are not located in the same processing point. On the other hand, the fact of processing the data using a common processing point implies the possibility of using a uniform logic, having a whole perspective about the data collecting and processing strategy previously not available. Besides, this approach would increase the data traffic on networks due to all the sensors should inform continuously the data to be processed, consuming an important volume of resources and incorporating a delay associated with the data travel time between the sensor and the processing unit.

A balance between consumed resources, processing pertinence, business requirements, processing, and communication available technology needs to be reached to determine the feasible point in which processing and storage need to be located. Thus, at the light of the global data processing strategy, the processing could be distributed as near to sensors as possible, or by the contrary, to be located at a common processing point. Also, when the decision is to distribute the processing, it could be located in some intermediary point between sensors and the main processing unit, or alternatively, as near to sources as possible. Here, these alternatives are discussed to contrast the strengths and weaknesses of each one.

### 6.5.1 A centralized data processing approach

This kind of approach consists of a central unit able to process data in an integrated way, collecting data from different data sources. The application's logic is integrated jointly with the processing unit, sharing the available resources for carrying forward their functionalities.

In this kind of environment, sensors are cheap, accessible, available, in limited capacity, and easily replaceable because the main function is to collect data itself. A priori, it sounds like something good, but if it is necessary to make some kind of local processing in the sensor, it is really a limitation. In this sense, it is important to design the primary use of the sensors and the expected role in relation to the whole data collecting strategy. Because sensors have limited resources in this architecture, they will transmit all the data as soon as they obtain each value, increasing the transmitted data volume. In case of some inconvenience with the network, this kind of devices will discard data due to inadequate processing and storage capacity.

Each sensor is independent of the processing unit, unknowing any aspect of the global processing logic. Each one only sees its immediate collecting environment, limiting its behavior to the local data gathering, and even not always is autonomous. This implies that the sensor itself not always could collect data by its own media, but it should be linked with another device that provides energy. For example, the DS18b20 sensor is able to measure temperature but if and only if it is connected to some processing board such as Arduino One [42].

On the one hand, the unification of the data processing provides a global perspective that allows articulating the application's logic with the data meaning spread out along the field. On the other hand, the complexity is increased due to all the collected data from heterogeneous data sensors are received in a unique point, being necessary the interpretation, some kind of conversion, among other aspects depending on the type of sensors and associated data formats. Thus, those additional functionalities are absorbed by the central processing unit, increasing the global processing time.

### 6.5.2 A distributed data processing approach

In this perspective, all or a part of the application's logic is distributed among all the components that integrate the processing architecture. The processing unit is not located at one given point, on the contrary; it is spread out along the components that integrate it. This allows distributing the business logic carrying it as near as the source as it is possible and necessary. On the one hand, it decreases the transmitted data volume due to part of the processing happens near to the collecting device. On the other hand, it increases the requirements of coordination in the collecting data strategy for avoiding risks of isolations.

Because each data collector requires a certain level of autonomy and processing ability, this incorporates some kind of local buffer and the possibility of data processing. For example, in this architecture, it is possible to find equipment such as Arduino Mega or Raspberry Pi articulating other sensors but acting as responsible for the data collecting from all the connected sensors. That is to say, as a difference with the previous architecture, in this case, the concentrator device is responsible for informing and processing the measures of a set of sensors (i.e., The Arduino or Raspberry Pi) [43].

This possibility of incorporating storage and processing capacities in this type of sensors' concentrator device allows providing partial results near to the data source, being able to detect early different risks depending on the monitoring requirements (e.g., fire and flood).

One of the challenges in this distributed environment is how the application's logic is distributed and balanced among components, discriminating each played role (e.g., collector, gateway, processor, and analyzer). Another no minor aspect falls into the heterogeneity associated with the components, which will affect directly on how the level of workload is shared out. It is important to highlight that even when the data volume related to measures is decreased due to the local processing, the processing overhead related to the coordination and the provision of partial results are increased.

Figure 6.3 synthesizes the general perspective of the data processing from the collecting point of view. Figure 6.3(a) represents a centralized collecting behavior, where sensors play a passive role in the sense that they only provide data. The communication is established directly between sensors and the central processing unit (CPU), which eventually could be partially or totally virtualized employing cloud resources. In this scenario, users send queries to the CPU to obtain updated data.

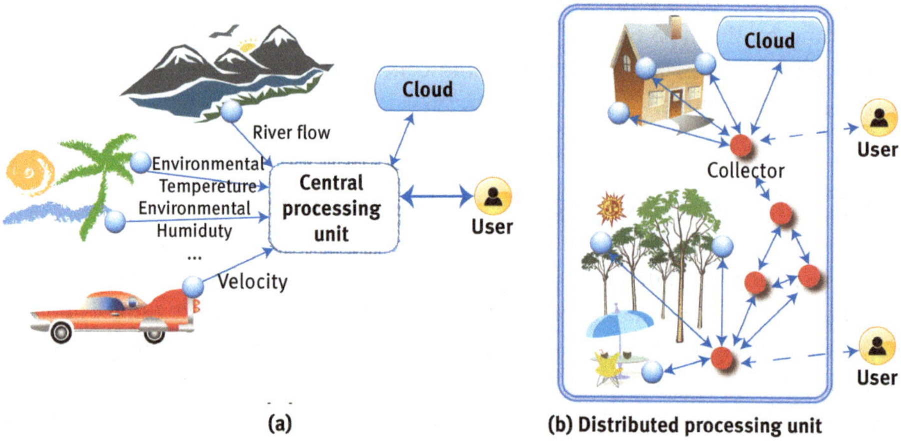


**Figure 6.3:** A general perspective of the data processing from the collecting point of view.

On the other hand, Figure 6.3(b) schematizes a distributed collecting behavior, where sensors are communicated through a collector. The collector is a component with a certain level of storage and processing, able to work in a collaborative and distributed way with other collectors. Thus, the collecting strategy is globally distributed, keeping a certain level of autonomy in each collector. In this case, sensors can interact with collectors, deploying an active behavior and being part of the answer to users. Also, collectors could implement all or part of its functionalities using virtualized resources in the cloud.

Table 6.3 allows to complement Figure 6.2 establishing a comparative perspective between centralized and distributed data collecting strategies. In a centralized strategy the dataflow is unidirectional because the sensor only provides measures without interaction capacities, while in a distributed environment each collector interacts with sensors (e.g., reviewing the current status) being able to interchange its own measures or from third collectors (i.e., indirect transmission). As previously was said, the data processing and business logic are distributed due to the collectors have resources and abilities to collaborate, while is limited to one unique place in the centralized architecture. Thus, sensors in centralized architecture only provide measures, while in a distributed environment they are additionally able to interchange processing results, metadata (i.e., description about other data), and so on. Clearly, the sensor's behavior in a centralized environment is passive (i.e., such as data provider), while it is active in a distributed context. The possibility of answer queries is another distinctive aspect in where the distributed environment is able to provide approximated data using collectors, while the centered context only answers the CPU but with full visibility of data. Finally, because the collectors have certain autonomy, they are responsible for obtaining and informing data (e.g., in case of missing value, it could detect and fix the situation discarding the miscalibration or estimating a value), while in the centered context the only responsible for data are the sensors. In this way, it is important to mention that the sensors involved in a centralized data collecting strategy are cheaper than the distributed environment due to the necessity of articulation and autonomy of the last one, which provokes that its budget is increased.

**Table 6.3:** Synthesis comparative between a centralized and distributed data collecting architectures.

| **Concept** | **Centralized** | **Distributed** |
|---|---|---|
| Kind of dataflow | Unidirectional | Bidirectional |
| Data processing and business logic | Centralized | Distributed |
| Kind of interchanged data from sensors | Only measures | Measures, results, metadata, etc. |
| Behavior | Passive sensors. Monitoring without interaction | Active behavior. Sensors and collectors |
| Data transmission | Direct | Direct and indirect |
| Query | Central processing unit | Collectors in a collaborative way |
| Data responsibility | Each sensor | Collectors |
| Cost | Cheap sensors (low) | Sensors articulated with collecting devices (more expensive than centralized) |

The decision about the kind of architecture to be used in a data collecting strategy will be based on a set of requirements. In this aspect, it is important to mention that there is not an ideal architecture for all the environments, but there are different architectures that could fit better or not to certain contexts depending on the necessities. The essential aspect in a data collecting strategy is clearly to know what the entity under monitoring is, the characteristics being monitored, the eventual relationship between characteristics, the way in which measures are obtained, and how to interpret each value under given scenarios and entity's states. Said in another way, the data collecting strategy should be supported by a measurement framework and all the components (i.e., sensors, collectors, etc.) be aligned with it.

## 6.6 Modeling the data stream's content

In general terms, a data stream could be represented as an unbounded sequence of data. However, the grey zone in this definition is related to the origin and the meaning of data. On the one hand, the common assumption is that the unbounded sequence of data is originated from one unique data source over which there is no control or some way of influencing. On the other hand, the second assumption is related to the meaning of data, that is to say, the data could be one unique kind of value provided over time with its corresponding variations, or alternatively, it could be understood such as a data structure with certain structure (i.e., a tuple) continuously provided over time.

Both the origin such as the provided data structure have a special interest in any measurement process because the original data structure affects directly the traceability and reliability, while the provided data structure has a direct influence on the content itself and the aim of the measurement process. For that reason, acquiring an agreement around these two aspects constitutes a key asset to foster the repeatability, extensibility, and consistency of any quantification process that needs to interpret values at the light of indicators using a set of given decision criteria.

This kind of discrimination between the origin (or data source) and the data structure associated with the unbounded sequence allows to introduce the idea of cooperative and exclusive data streams that are explained accordingly in the following sections to model each alternative.

### 6.6.1 The exclusive data streams

An exclusive data stream is defined as *an unbounded sequence of domain-known and atomic data with an autonomous, simple, and independent data source.*

When definition refers to a sequence, it indicates a list of data ordered, following a timestamp that represents the order in which data were generated. Thus, the ordering has a relationship to the generation time itself that simultaneously depends on the data source and the monitored event. This is important to highlight because the sequence implies the existence of some event jointly with a device (virtual or not) actively monitoring it. In other words, the underlying idea related to the "sequence" is to receive continuously the updated situation related to an event, and not one sporadic data.

The indication related to "unbounded" makes focuses on the data volume, without previously establishing any limitation to it. Thus, the idea is to provide as much data as necessary about an event, the data generation has not a given bound, and the data rate depends on the data source. This is important because it is not possible to anticipate a data size, or even, predict the rhythm in which they will arrive. Even more, such rhythm could be absolutely variable, implying that the data rates go changing over time without any predefined periodicity.

The atomic data is associated with the representation of a single value for a unique concept or meaning under analysis, being this immutable. Thus, the meaning that is being monitored does not change over time, for example, if the device measures the environmental temperature, only such concept will be informed using this stream. The atomicity is similar to the relational model, and it implies that only one single value is communicated by each timestamp. It could be associated with a number, a position, bytes, among others but always one single value at the end.

The kind of value domain related to each informed single value is previously known and it is immutable. That is to say, the defined data type jointly with the set of constraints is established to bind the normal and expected values' range. Any intents of changing the value domain (i.e., data type or constraints) associated with a data stream will imply a redefinition of it.

Following the definition, the data source should be autonomous, simple and independent: (1) By autonomous, it is indicated that it should be able to keep running independently of who is reading its values. The collection method is dependent on the own device; (2) By simple it is understood that each value has a unique origin that affects in a direct way the device traceability; and finally (3) By independent it indicates that it is not influenced by external sources.

An attribute represents a concept to be quantified related to some concept or event that needs to be monitored. Equation. (6.1) defines the set "$A$" as a set of potential attributes:

$$A = \{a \mid a \text{ is an attribute}\} \tag{6.1}$$

Equation (6.2) describes the way in which a value is obtained from an attribute. That is to say, given an attribute, it is possible to define a method by means of a function able to get a value (e.g., a measure). The set "$M$" indicates all the values obtained by the application of the function over a given attribute. The kind of results will depend on the type of defined function and the used method:

$$M = \{m \in D | m = f(a), \forall a \in A\} \tag{6.2}$$

Equation (6.3) formally defines an *exclusive positional data stream*, representing that the unbounded sequence of values on an attribute "*a*" is ordered based on the arriving (i.e., a certain value of "*i*") of each obtained value, employing the defined function in eq. (6.2). However, this stream contains single values without any temporal stamp. The notion of order is limited to the position, assuming that the element $m_i$ is previous to $m_{i+1}$ and that way successively:

$$\forall a \in A, i \in \mathbb{N} / S^p_{ex} = \{f(a)_i, f(a)_{i+1}, f(a)_{i+2}, \ldots\} = \{m_i, m_{i+1}, m_{i+2}, \ldots\} \tag{6.3}$$

Depending on the value domain "*D*" defined in eq. (6.2), it is possible to deal with a numerical, categorical, or ordinal data stream. An exclusive numerical positional data stream contains only numerical values. An exclusive categorical positional data stream contains categorical values as text. Finally, an exclusive ordinal positional data stream contains ordinal values. The ordinal values could be represented as a numerical set where the value represents a given order, not necessarily a magnitude, or alternatively, it could be represented as text.

For example, it supposes that the chosen attribute is the corporal temperature of a person and using some device, it is possible to obtain a value, an exclusive numerical positional data stream as in eq. (6.4):

$$S^p_{ex}(Corporal\ Temperature) = \{36.0,\ 36.1,\ 36.1,\ 36.08,\ 36.07,\ 36.06,\ \ldots\} \tag{6.4}$$

Subtly different is an exclusive categorical positional data stream related to the sensing of colors as a text, as expressed eq. (6.5):

$$S^p_{ex}(colour) = \{Red,\ Blue,\ Yellow,\ Red,\ Green,\ Pink,\ \ldots\} \tag{6.5}$$

Following the analysis, an exclusive ordinal positional data stream could be represented by means of text (see eq. (6.6)) or numbers (see eq. (6.7)). For example, it supposes that the value of the temperature is interpreted by an indicator informing in a continuous way one of the following values: 1. Hypothermia, 2. Low Temperature, 3. Normal, 4. Fever, and 5. Very high fever:

$$S^p_{ex}(corporal\ temperature) = \{Normal,\ Fever,\ Fever,\ Fever, Very\ High\ Fever,\ \ldots\} \tag{6.6}$$

$$S^p_{ex}(corporal\ temperature) = \{\ 3,\ 4,\ 4,\ 4, 5,\ \ldots\} \tag{6.7}$$

Equation (6.8) defines an *exclusive temporal data stream* that describes an unbounded sequence of values based on an attribute "*a*" jointly with a timestamp (i.e., "*t*"). Both value and timestamp correspond with the monitored attribute, that is to say, the value $m_i$ is obtained at the time $t_i$ for the attribute "$a$" and they need to be view as an ordered pair or vector. An important aspect in this kind of data stream is the mandatory existence of the timestamp, that is to say, given an ordered pair $(m_i, t_i)$ the value of $m_i$ could be missing but the timestamp must be always present, else the ordered pair is not considered because that directly would affect the order's principle indicated in the following equation:

$$\begin{aligned}\forall a \in A, t \in T, i \in \mathbb{N} / S_{ex}^{t} &= \{ (f(a)_i, t_i), (f(a)_{i+1}, t_{i+1}), (f(a)_{i+2}, t_{i+2}), \ldots \} \\ &= \{(m_i, t_i), (m_{i+1}, t_{i+1}), (m_{i+2}, t_{i+2}), \ldots \}\end{aligned} \tag{6.8}$$

Similarly, depending on the kind of definition in the set "$D$" in eq. (6.2), the exclusive temporal data stream could be derived in a numerical, ordinal, or categorical data stream.

Particularly, the ordering constitutes an important aspect in measurement projects where the order in which each value arrives could be determinant depending on the kind of analysis carried out, for example, to determine whether a person has fever or not. In case some project does not require specification about the order, eqs. (6.8) and (6.3) could be unified extracting the component that defines the order (be it the position or timestamp). Thus, an *exclusive data stream* is an unbounded sequence of single values on an attribute "$a$," employing the defined function in eq. (6.2) in a successive way (represented it in eq. (6.9) through the accumulation of the apostrophe character) without that necessarily implies a given order for the processing:

$$\forall a \in A / \; S_{ex} = \{f(a), f(a)', \; f(a)'', \ldots\} = \{m_i, \; m'_i, m''_i, \ldots\} \tag{6.9}$$

In other kinds of data models, such as the relational or columnar data model, the scope of the set is limited, which implies that each one could be as bigger as be necessary, but they have a size at the end. These sets could be processed through different operators that could be synthetically indicated as blocking or nonblocking. A nonblocking operation (e.g., a projection in the relation algebra) allows continuing the data processing without incorporated delay. However, a blocking operation (e.g., an "order by" in SQL) does not provide a result till the operation reaches the end. In other words, the data processing does not continue till the result of blocking operation is available to be incorporated as an input of the next operation inside an execution plan.

Up to here, it was incorporated the positional exclusive data stream and the temporal exclusive data stream. In each kind and depending on the defined domain, variations related to numerical, ordinal or categorical values could be present.

It is known that a way to process a data subset from a data stream is by means of windows [44, 45]. The window represents a finite subset of data created by applications of restrictions over an unbounded data stream.

Because the positional data streams does not have a concept of time associated, we define the concept of the arriving time as the instant in which data arrives at the processor unit, without it has a relationship with the time in which the data was generated. The arriving time (i.e., "*at*" in eq. (6.10)) depends on the time where the processing unit has its first contact with the datum, and that it is independent of the monitored attribute or the kind of received value:

$$\forall s \in S^{p}_{ex} \wedge i \in \mathbb{N}/t = at(s_i) \wedge t \leq RTS \tag{6.10}$$

Basically, given a set of positional exclusive streams and "*s*" a stream belonging to it, the arriving time of an "*i*" element to the processing unit is known using at ($s_i$), obtaining a timestamp that will be equal or lesser than a reference timestamp (RTS) of the local clock. In this sense, it is worthy to mention that the timestamp is not associated with the data generation time but the arriving time at the processing unit.

Thus, a *physical window* is time-based and its definition for exclusive data stream is synthesized in eqs. (6.11) and (6.12):

$$\forall s \in S^{p}_{ex} \exists w_T \subset s/w_T\text{: } RTS - wishedTime \leq at(s_i) \leq RTS \tag{6.11}$$

$$\forall s \in S^{t}_{ex} \exists w_T \subset s/w_T\text{: } RTS - wishedTime \leq t_i \leq RTS \tag{6.12}$$

Equation (6.11) establishes that the temporal window "$w_T$" (a subset of "*s*") will contain data values that have arrived at the processing unit between RTS and RTS-*wishedTime*, in where "*wishedTime*" is a relative temporal magnitude (e.g., 1 min). On the contrary, eq. (6.12) defines that the window "$w_T$" will contain data values in which their generation timestamp falls in the interval [CTS-*wishedTime*; CTS].

The *logical window* is data volume based, which implies that a certain number of elements are retained based on a threshold (e.g., 1000). The number of elements inside a window is represented as the cardinality (e.g., |*w*|). As it is possible to appreciate in eqs. (6. 13) and (6.14), there is no structural difference in the definition, though the elements in each set are different (positional single values versus temporal time-based ordered pairs):

$$\forall s \in S^{p}_{ex} \exists w_L \subset s/\ w_L\text{:}|w_L| \leq Threshold \tag{6.13}$$

$$\forall s \in S^{t}_{ex} \exists w_L \subset s/w_L\text{:}|w_L| \leq Threshold \tag{6.14}$$

As it was introduced, the physical windows are based on the time, while the logical windows are limited by the data volume. However, it is possible to update the definition of sliding and landmark windows that refer to the way in which each window updates its content. The *sliding window* keeps the established limits, but it updates the extremes for replacing old elements by new elements. In case of physical windows, it is possible to indicate that the window needs to contain the last 5 min of data. Updating eqs. (6.11) and (6.12) with the current timestamp (CTS), it would be sliding windows as defined in eqs. (6.15) and (6.16):

$$\forall s \in S^p_{ex} \exists w_T \subset s/w_T\colon CTS - wishedTime \leq at(s_i) \leq CTS \tag{6.15}$$

$$\forall s \in S^t_{ex} \exists w_T \subset s/w_T\colon CTS - wishedTime \leq t_i \leq CTS \tag{6.16}$$

As a difference with eqs. (6.11) and (6.12), the lower and upper endpoints are variable because they are permanently updated with the current timestamp. By definition, data could be present in the window in a given moment, but it will be discarded only with the pass of the time. Of course, the retention criteria between eqs. (6.15) and (6.16) continue being different, because of the positional data stream that employees the data arriving time, while the temporal data stream uses the data generation time.

In the case of the logical windows, it is possible to define something like "LAST 1000." Thus, the window size follows being the same, but with each new arrival, new data is added, and the oldest data is discarded in order to keep updated its contents.

Alternatively, it is possible to define landmark windows like those in which one point (i.e., initial or final) that defines the window is updated against the occurrence of an event, being the data volume related to its content something variable. For example, the event could be the accident in a factory (i.e., milestone), each time that an accident happens, the window content could be restarted as it is described in eqs. (6.17) and (6.18):

$$\forall s \in S^p_{ex} \exists w_T \subset s/w_T\colon milestone \leq at(s_i) \leq CTS \tag{6.17}$$

$$\forall s \in S^t_{ex} \exists w_T \subset s/w_T\colon milestone \leq t_i \leq CTS \tag{6.18}$$

Figure 6.4 graphically describes the main perspectives related to exclusive data streams introduced previously through eqs. (6.1) to (6.18). The positional data streams correspond with single values that are organized based on their arrival, obtaining eventually the notion of time from the arriving time given by its processing unit. That is to say, the timestamp in each positional data stream corresponds with the instant in which the data processing unit has read the data. The temporal data streams, on the other hand, could be viewed as an ordered pair in which each pair is composed of the measure or value related to a concept jointly with the timestamp in which the value or measure has been taken from the data source. This is a key difference in terms of data traceability because the temporal data stream can establish a relationship between the data and the moment in which it was generated, on the contrary, the positional data stream has an artificial timestamp derived from its processing order.

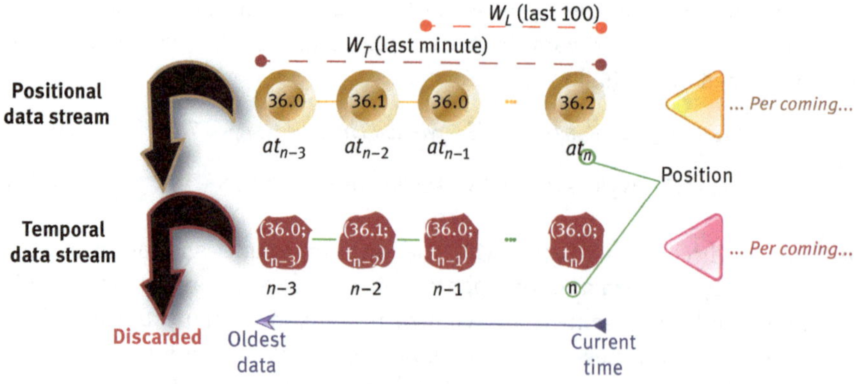


**Figure 6.4:** Perspectives from the exclusive data streams.

The notion of the subset of the data streams is modeled through the concept of windows. They could be physical or logical, updating their content in a sliding or landmark way. Anyway, data could belong to a given window during a certain period; however, the data will soon be discarded to be replaced by new data. This is part of the essence of the data streams, to keep the content as updated as possible in order to process or analyze the current or last known data for making decisions.

The data joining or matching operations could be carried out by mean of the data value, its position, or the given timestamps. Even, it is possible to process temporal and positional data streams together, only how the data will be crossed needs to be established. As a result of the chosen operation, a new data stream is created; however, the result may not necessarily be strictly temporal. That is to say, if the processing unit makes the matching of the data streams described in Figure 6.4 based on its position, the first result would be (36.0; (36.0; $t_o$)). The value of "$t_o$" indicates the generation time of the last one 36 but not of the first one transitively. The timestamp is exclusive of the associated datum and it is not transitive in relation to other data. In this way, the new data stream in this example will contain part of their data with the data generation timestamp coming from the data source, being able to incorporate the arriving time to the processing unit for the positional data (i.e., ((36.0; $at_0$);(36.0; $t_0$))).

## 6.6.2 The cooperative data streams

The underlying idea associated with the cooperative word is to use a common carrier to transport different concepts together, optimizing the use of resources and decreasing the idle time. The ideal situation for a data channel is to keep it as near as possible to 100 percent of capacity but avoiding the overflows. The overflow represents a situation in which the data production rate exceeds the channel capacity for data transmission, provoking the loosing of data in the data source.

A *cooperative* data stream is defined as *an unbounded sequence of composed data with a given immutable and well-known data structure associated with one or more autonomous, simple, and independent data sources gathered under the concept of a collector.*

The figure of the *collector* represents a component in which a set of data sources are fully communicated with a processing endpoint, using it as an intermediary. Data sources provide continuously values to the collector, while the collector defines the data transmission policy articulating the requirements of all the sources jointly. In other words, between data source and collector there are two roles differentiated. On the one hand, the data source is responsible for providing data (e.g., a measure, picture, and audio). On the other hand, the collector is responsible for locally storing the received data, fusing it in terms of the target data format keeping immutable its meaning, providing transport services to the processing endpoint, and provide approximate answers to users based on the local data (e.g., an edge device):

$$\forall a \in A, j \in \mathbb{N} / \vec{a} = \left(a_1, a_2, \ldots, a_j\right) \tag{6.19}$$

Equation (6.19) describes the data structure (i.e., DS) to be monitored through a cooperative data stream. Up to here, nothing has been said about the timestamp, the equation earlier is limited to describe the set of attributes to be informed, their order, and from the attribute definition, their value domains are known. Once the data structure has been defined, the data stream will not change it, otherwise the data stream definition would be changing:

$$\vec{M} = \{\exists \vec{m}_i | \vec{m}_i = \left(m_{a1}, m_{a2}, \ldots, m_{aj}\right) = f(\vec{a}) \ldots, \forall a \subset \vec{a} \wedge i, j \in \mathbb{N}\} \tag{6.20}$$

Equation (6.20) represents the set of valued vectors (i.e., $\vec{m}$) from the set of defined attributes (i.e., $\vec{a}$), that integrates the set of all the valued vectors known as $\vec{M}$.

Form eqs. (6.19), (6.20) and (6.3), it is possible to derive the definition of a *cooperative positional data stream* as an unbounded sequence of valued vectors from a set of attributes (i.e., $\vec{a}$), ordered based on its arriving:

$$\forall a \subset \vec{a}, i \in \mathbb{N} / S^p_{co} = \left\{f(\vec{a})_i, f(\vec{a})_{i+1}, f(\vec{a})_{i+2}, \ldots\right\} = \{\vec{m}_i, \vec{m}_{i+1}, \vec{m}_{i+2}, \ldots\} \tag{6.21}$$

Equation (6.21) describes the formal definition for the cooperative positional data stream. In such sense, each "$\vec{a}$" could be considered as a tuple in analogy to the relational model. In this way, the cooperative positional data stream could be described as an unbounded sequence of tuples ordered in accordance with its arrival.

It is worthy to mention that there is no warranty that each element at the vector will be always valued, that is to say, it is possible that some attribute "$a$" in a given time, does not have a value. This will imply that some attributes in the vector have missing values. However, as it was indicated in eq. (6.20), the vector must exist, which implies that at least one value "$m_{ai}$" must be available. In effect, a vector with all its attributes without value does not represent any received data, and it would consume resources without apparent necessity:

$$\begin{aligned}\forall a \subset \vec{a}, t \in T, i \in \mathbb{N} / S^t_{co} &= \left\{ \left(f(\vec{a})_i, t_i\right), \left(f(\vec{a})_{i+1}, t_{i+1}\right), \left(f(\vec{a})_{i+2}, t_{i+2}\right), \ldots \right\} \\ &= \left\{ (\vec{m}_i, t_i), (\vec{m}_{i+1}, t_{i+1}), (\vec{m}_{i+2}, t_{i+2}), \ldots \right\}\end{aligned} \tag{6.22}$$

A *temporal cooperative data stream* could be defined as an unbounded sequence of immutable vectors of attributes simultaneously valued and ordered based on its collecting timestamp. In this sense, the data structure related to each vector is fixed, and it cannot be changed over time. As it was said before, each vector at least must contain a value for an attribute, even when the rest of the attributes that compose the vector "$\vec{a}$" have no value. However, eq. (6.22) defines the temporal cooperative data stream with a strong assumption that all the values of attributes have been provided at the same instant. This is important to highlight because the cooperative streams are provided through collectors, and each collector is linked a set of sensors. In this way, the collector must warranty that the informed values through this kind of data stream correspond to the same timestamp.

Because each attribute "$a$" could be quantitative, ordinal, or categorical, each vector could be integrated of a combination of such kinds of attributes, for such reason, these streams are known as *mix temporal cooperative data streams*, or simply temporal cooperative data streams because of the common situation. In case of all attributes of a vector "$\vec{a}$" be of the same kind, the data streams could be indicated as: (1) *Categorical temporal cooperative data stream:* when all the attributes that compose it are nominal or categorical; (2) *Ordinal Temporal Cooperative data stream:* when all the attributes that compose it are ordinal; and (3) *Quantitative Temporal Cooperative data stream:* when all the attributes that compose it are quantitative.

The concept of the *physical window* could be extended based on eqs. (6.11) and (6.12) as shown in eqs. (6.23) and (6.24). It is possible to appreciate that the main change is associated with the structure of the data stream and not about the timestamp:

$$\forall s \in S^p_{co} \exists w_T \subset s / w_T\text{: } RTS - wishedTime \leq at(s_i) \leq RTS \tag{6.23}$$

$$\forall s \in S^t_{co} \exists w_T \subset s / w_T\text{: } RTS - wishedTime \leq t_i \leq RTS \tag{6.24}$$

In the same way for the *logical windows*, they could be extended based on eqs. (6.13) and (6.14) as shown in eqs. (6.25) and (6.26):

$$\forall s \in S^p_{co} \exists w_L \subset s / w_L\text{:} |w_L| \leq Threshold \tag{6.25}$$

$$\forall s \in S^t_{co} \exists w_L \subset s / w_L\text{:} |w_L| \leq Threshold \tag{6.26}$$

It is important to highlight that the cardinality given in eqs. (6.25) and (6.26) corresponds with the number of contained vectors, independently the number of attributes in each vector.

Something similar happens in relation to the sliding and landmark windows. Based on eqs. (6.15) to (6.18), *sliding windows* are specified by eqs. (6.27) and (6.28) for positional and temporal cooperative data streams respectively. Thus, *landmark windows* are described by eqs. (6.29) and (6.30) accordingly:

$$\forall s \in S^p_{co} \exists w_T \subset s/w_T\text{: } CTS - wishedTime \leq at(s_i) \leq CTS \quad (6.27)$$

$$\forall s \in S^t_{co} \exists w_T \subset s/w_T\text{: } CTS - wishedTime \leq t_i \leq CTS \quad (6.28)$$

$$\forall s \in S^p_{co} \exists w_T \subset s/w_T\text{: } milestone \leq at(s_i) \leq CTS \quad (6.29)$$

$$\forall s \in S^t_{co} \exists w_T \subset s/w_T\text{: } milestone \leq t_i \leq CTS \quad (6.30)$$

The data joining or matching operations could be carried out by the mean of the data value, its position, or the given timestamps contained in an attribute. For example, algorithms such as the Symmetric Hash Join could be employed [46]. As a difference with the previous exclusive data streams, the cooperative data streams contain a set of attributes between which it is possible to choose one or more, through which the joint operation could be carried out. Thus, each cooperative stream (be it positional or temporal) could be analyzed as an unbounded sequence of tuples, simulating each data stream as an infinite bidimensional data organization (e.g., a table).

However, it is important to mention that the matching done is not dependent of the relational model; it only represents an interpretation by an analogy.

The matching or joining operations are not exclusive of the cooperative data streams, which implies that a joining operation between a positional exclusive data stream and a temporal cooperative data stream could be carried out in the measure that the crossing criteria are defined. Once defined, the result will be a cooperative data stream depending on the number of projected attributes.

In this way, the operations defined over the data streams allow giving as result new data streams derived from those employed as the origins, following some predefined criteria or restrictions. In other words, the cooperative and exclusive data streams could be interpreted from the bidimensional point of view and the previous works on bidimensional data models could be reused (e.g., bitemporal model, relational model, columnar model) [47–50].

The contrast between Figures 6.5 and 6.4 allows visually to see the differences between cooperative and exclusive data streams. As illustrated in Figure 6.5, each data item in a cooperative data stream has a structure determined by an ordered sequence of attributes (i.e., $a_i$). Such an ordered sequence could be known as a vector or tuple in an analogy with math or relational model respectively. But in the temporal cooperative data streams, all the attribute values simultaneously depend on the same timestamp ($t_{n-1}$ in Figure 6.5). This is an important difference of the

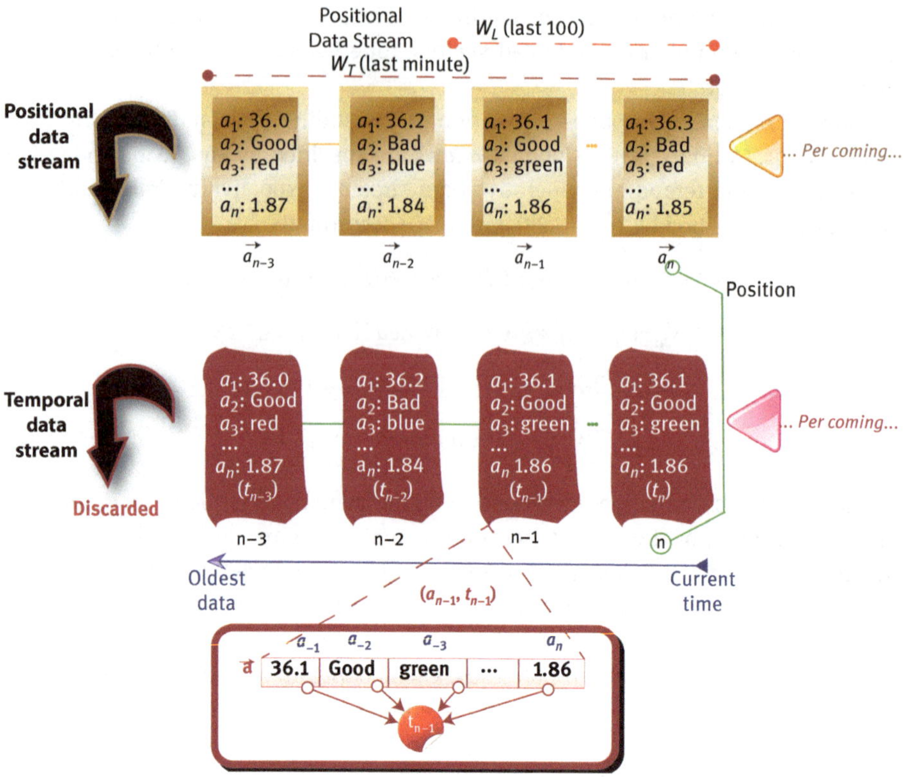


**Figure 6.5:** Perspectives from the cooperative data streams.

exclusive data stream because the last one has a timestamp specifically associated with its value, while the cooperative data stream is shared along with all the elements that compose the vector $\vec{a}$.

Positional and temporal cooperative data streams have time differences in, while the first one imposes its time by the arrival time, the second one brings the timestamp from the data source as part of their data. In both cases, always there will be new data arriving for replacing the oldest data, being the last ones discarded by obsolescence.

The concept of logical and temporal windows has a behavior similar to the exclusive data streams, but the structure related to each data item is subtly different as it can be appreciated in Figure 6.5. Also, the landmark and sliding windows could be extended for being used with cooperative data streams.

In this scenario, one could think that there is not space for semistructured data or unstructured data that would be a mistake. Why? Because from eqs. (6.2) and (6.20), the set $M$ is defined as a valuation of certain attribute "$a$" giving a result known as "$m$," which it would be informed into a data stream. It is not necessary that "$m$" must be an atomic value in terms of the relational model. In other words, "*m*" represents the bucket in which data arrives to be processed and the structure of such data will be in terms of its value domain. Thus, "m" could be an XML file generated from a given data source or data collector, which needs to communicate it to the data processing unit. Thus, it is possible to see the exclusive data streams as a sequence of buckets (be it temporal or not), while the cooperative data streams as a sequence of "rows" composed of buckets in which in each position could have different domains.

As it is possible to appreciate in Figure 6.6, the main distinctive characteristics that allow adequate discriminating between data streams could be synthesized around the timestamp, timestamp origin, data structure, number of associated attributes, data order, use of intermediaries, and the supporting of windows for its data processing.

## 6.7 Operating over exclusive and cooperative data streams

With the aim of analyzing the exclusive and cooperative data streams at the light of basic operations, they are organized from two points of view. On the one hand, operations related to the set theory, that is to say, the union, intersection, difference, and cartesian product. On the other hand, operations associated with the projection, restriction, joining, and division.

For better organization, this section introduces a notation to be used for expressing the mentioned operations. Later, each operation will be individually introduced jointly with its exemplification.

### 6.7.1 Notation

Figure 6.7 synthesizes the proposed idea in relation to the notation for exclusive and temporal data streams. The differentiation between exclusive and cooperative is derived from the number of attributes that compose it. That is to say, when the number of attributes between brackets is upper or equal to two, the data stream is cooperative, while when the number of attributes specified between brackets is just one, the data stream is exclusive.

As it is possible to appreciate in the mentioned figure, the positional and temporal aspects are separated from the list of attributes. This is due to both aspects have no dependency on the attributes or their contents, they are independent of their values. For example, if it is necessary to reach values of certain attributes between positions 2 and 5, that kind of restriction will have no relationship with the contained values in those attributes but with the indicated position.

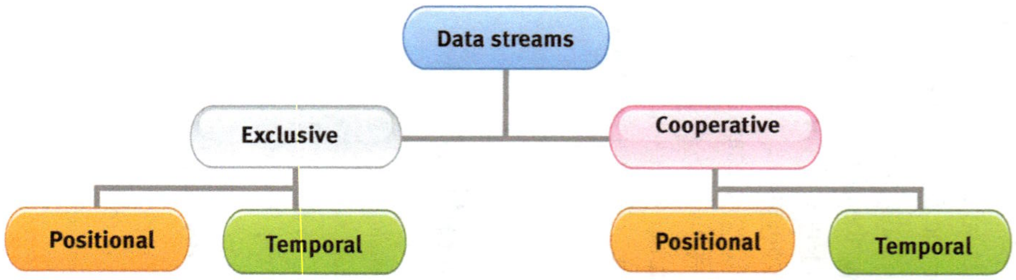


| | Exclusive – Positional | Exclusive – Temporal | Cooperative – Positional | Cooperative – Temporal |
|---|---|---|---|---|
| Timestamp | Artificial | Intrinsic | Artificial - shared | Intrinsic - shared |
| Timestamp origin | Arrival time | Data source | Arrival time | Collector |
| Strtucture | $a$ | $(a, t_i)$ | $a$ | $(a, t_i)$ |
| Number of attributes | one | One + timestamp | “$n$” | “$n$” + timestamp |
| Order based on the | Arrival position | Timestamp | Arrival position | Shared timestamp |
| Intermediary | Arrival position | No | Yes (collector) | Yes (collector) |
| Logical windows | ✓ | ✓ | ✓ | ✓ |
| Physical windows | ✓ | ✓ | ✓ | ✓ |
| Sliding windows | ✓ | ✓ | ✓ | ✓ |
| Landmark windows | ✓ | ✓ | ✓ | ✓ |

**Figure 6.6:** Comparative synthesis of the main characteristics of the kinds of data streams.

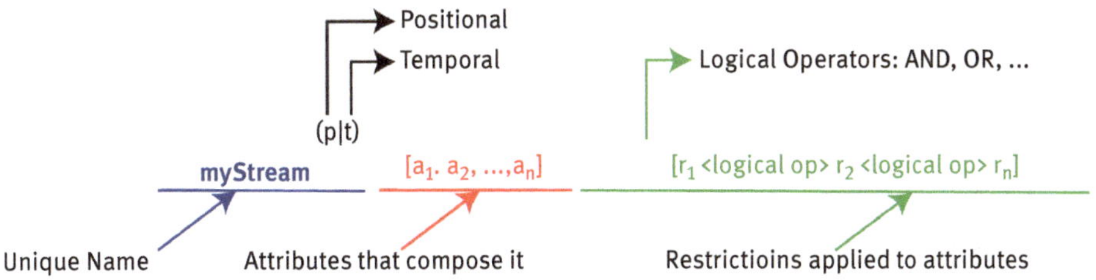


**Figure 6.7:** Notation for temporal and positional data streams.

For a better description of this proposed notation, Table 6.4 describes some examples jointly with its description.

An alias establishes a synonym that allows to manage different expressions in a friendly way. The way in which an alias is defined is similar to the Structured Query Language (SQL), employing the reserved word "as". For example, it would be a good idea to define an alias for obtaining the last ten elements from the positional data stream as it is shown in eq. (6.31):

$$myPos^{p}[*].p \ between \ last() - 10 \ and \ last() \ AS \ last10 \tag{6.31}$$

The "*" indicates all the attributes defined in the data stream. Thus, the last ten data items arriving through the $myPos^{p}$ data stream will be informed with all the attributes under the synonym named "*last10*". When the expression last10[colour] is present, it really represents the last ten colors arriving by the data stream $myPos^{p}$. An alias could be part of a new alias definition, for example, *last10[colours] as myLast10Colours*. The underlying idea is to simplify the involved expressions related to the data communication.

## 6.7.2 Operations

The *projection* operation consists in extracting a subset of attributes defined on a data stream and exporting them as a new data stream:

$$newDS = measures^{t}[dataSource, \ value, collector] \tag{6.32}$$

Equation (6.32) creates a new temporal data stream (i.e., the origin is completely temporal) named *newDS* that is composed of three attributes known as *dataSource*, *value*, and *collector*. Even, when it is possible that the data source named "measures" has more than three attributes, only the indicated attributes between square brackets will be projected:

$$measures^{t}[*](weight < 19 \ and \ colour = blue) \tag{6.33}$$

**Table 6.4:** Synthesis comparative between a centralized and distributed data collecting architectures.

| Notation | Description |
|---|---|
| $myStream^{t}[colour,\ height]$ | A temporal data stream named "stream," which contains two attributes defined as color and height. |
| $myStream^{t}.t\ between\ now() - 5secs\ and\ now()$ | The temporal stream restricts the contained data items to the last five seconds, independently the values contained in the attributes color and height. |
| $myStream^{t}[colour](colour = \text{blueORheightbetween5and10})$ | The temporal stream restricts the contained data items to those that contain the value "blue" for the attribute color, or alternatively those in which the height is between 5 and 10. As a result of restrictions, the attribute "colour" is informed. |
| $myStream^{t}[colour](colour = \text{blueORheightbetween5and10}).tbetween\ now() - 5secs\ and\ now()$ | Similar to the previous one but limiting the restrictions to the data items contained in the last 5 s. |
| $myPos^{p}[colour,\ weight]$ | A positional data stream named "myPos," which contains two attributes defines as color and weight. |
| $myPos^{p}.p\ beetween\ 5\ and\ 8$ | A positional data stream that continuously informs the data items located between positions 5 and 8 included. |
| $myPos^{p}[colour](weight\ between\ 5\ and\ 7).p\ between\ last() - 5\ and\ last()$ | It informs the last 5 positions for the attribute "colour" but only for those in which weight is between 5 and 7. |

The *restriction* operation limits the number of data items in the data stream, following a logical expression that will return TRUE or FALSE. When the expression returns TRUE, the data item is retained and informed in the result, otherwise, it is not considered. Figure 6.33 shows an example in which all the attributes related to the data stream named "measures" are projected but only to those data items that simultaneously have a weight lower than 19 and their color is blue.

Traditionally, the *cartesian product* operation realizes a combination of elements, combining them all against all. The challenge in this context is that given two data streams, neither has a limit nor size, and this is the reason for which the combination of all against all is relative because the stream's end is never known. An alternative for implementing it is through the windows, in which certain limits or sizes to each set could be established:

$$measures^{t}[a,\ b,\ c,e,f,g] = measures\_a^{t}[a,\ b,\ c] \cup measures\_b^{t}[e,f,g]] \tag{6.34}$$

$$measures^{t}[a,\ b,\ c,e,f,g] = measures\_a^{t}[a,\ b,\ c] \cup measures\_b^{p}[e,f,g]] \tag{6.35}$$

The *union* operation generates a new data stream integrated by all the data items and attributes coming from the data sources that give origin to it (see eq. (6.34)). The challenge in this kind of operation has a relation to the managing of temporality. That is to say, given two temporal data streams, the union will order the new data stream based on their respective temporal dimension. However, when two structurally different data streams (i.e., one positional and one temporal) need to be united, the new data stream will order data, mixing arriving, and generation timestamps. As you can appreciate in eq. (6.35), once the new data stream has been created, the data items will be considered as temporal and assuming the arriving timestamp as a generation timestamp.

The *intersection* operation generates a new data stream integrated by those data items that are present in both streams. Analogously to the cartesian product, this operation requires a previous knowledge about the volume of involved data. This could be implemented using windows based on data streams structurally compatible; otherwise, the result would be empty. The underlying idea around the concept "structurally compatible" is that at least there exists one common attribute between attributes and based on that, the common values between the common attributes will integrate the result.

The *difference* operation has the same limitations that the intersection and cartesian product operations, they need to have a limited set in a context in which each data stream is unbounded. That is to say, given an expression such as A-B the idea is that in the result will be all the data belonging to the set "A" not present in the set "B". The adopted strategy for carrying forward these operations is through the employing of windows:

$$measures^{p}[a,\ b,\ c,f,g] = measures\_a^{p}[a,\ b,\ c]\ natural\ join\ measures\_b^{p}[a,f,g] \tag{6.36}$$

$$measures^t[a, b, c, d, f, g] = measures_a{}^t[a, b, c] \text{ as } ma \text{ inner join } measures_b{}^t[d, f, g] \text{ as } mb \text{ on } (ma[a] = mb[d]) \quad (6.37)$$

The natural join operation creates a new data stream based on other data streams following established criteria. For example, on the one hand, eq. (6.36) creates a new positional data stream composed of the attributes "*a, b, c, f,* and *g*," matching the values for the attributes with the same name (i.e., attribute "*a*"). On the other hand, eq. (6.37) describes a new data stream generated through an inner join, matching the attributes "*a*" and "*d*" using aliases:

$$measures^t[a] = measures_a{}^t[a, c] / measures_b{}^t[c] \quad (6.38)$$

The *division* operation is between one cooperative and one exclusive data stream. The binary cooperative data stream has two attributes and it is compared with the available values in the exclusive data stream, each matched value will imply a new data item to be included in the result. The result will be integrated by the attribute of the cooperative data stream not present in the exclusive data stream. Equation (6.38) shows an example dividing a temporal cooperative data stream with a temporal exclusive data stream, the common attribute is "*c*" reason for which the result will be integrated of the attribute "*a*." Each data item containing the attribute "*a*" emerges from those data items in which the value for the attribute "*c*" is coincident between both data streams. It is worthy to mention that the exclusive data stream needs to be implemented using windows (i.e., divisor) because a finite size is required in order to contrast values coming from unbounded data streams (i.e., dividend) with its divisor.

Figure 6.8 represents an example of the division operation between one cooperative and one exclusive data stream for eq. (6.38). The number 5.0 in the result appears because the same value contains for the attribute "*c*" the values "ds1" and "ds2" corresponding to the divisor's values.

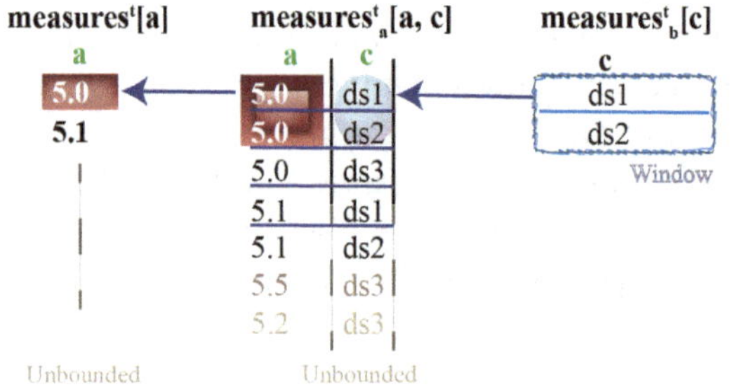


**Figure 6.8:** An example of the division operation between the cooperative and exclusive data streams.

## 6.8 About the overhead related to the processing perspective of the exclusive and cooperative data streams

For analyzing the overhead related to the processing perspective of the exclusive and cooperative data streams, a new library named *streamCE* was proposed as a proof of concept. It was developed in Java and released under the terms of the Apache 2.0 General Agreement License. The source code is freely available on Github, introducing a basic implementation of exclusive and cooperative data streams jointly with the implementation of the operations of Union and projection.[1]

To analyze the involved times associated with each operation, two simulations were carried out on the same hardware. The host was a MacBook Pro 2017 with Mac OS Catalina 10.15.1, Processor Intel I7 with four cores, 16 GB 2133 MHz LPDDR3, and Radeon Pro 560 4 GB Graphic Board.

The first simulation was carried out on the operation of the union. Two cooperative data streams were defined as inputs: *streamA* and *streamB*. The *streamA* had the numerical attributes named "*a*," "*b*," and "*c*," while the stream had the numerical attributes named "*d*," "*e*," "*f*," "*g*," "*h*," and "*i*." Each attribute was randomly filled avoiding null values to analyze the full required processing time. The result of the union operation was modeled as a cooperative data stream, taking the timestamp from the most recently updated data stream. In case one of the data streams is updated (i.e., *streamA* or *stream B*), the union took the updated value jointly with the last known data item from the second data stream, representing the situation in which one of them could have been interrupted.

The simulation created the data streams, after that the operator was created and linked to the data streams acting as inputs (i.e., *streamA* and *streamB*). Once the operator and input were established, a set of threads was created for acting on the data streams, producing the random values for each attribute. The processing time was analyzed continuously for 10 min. In this sense, the time consumed for the garbage collector had a huge impact on the unitary processing rate.

Figure 6.9 describes the mentioned unitary processing time throughout the 10 min of continuous processing. The peaks are related to the consumed time of the garbage collector, which incorporates important jumps in relation to the general unitary processing time. On the one hand, the unitary processing rate tends to be under 0.01 ms, even reaching rates such as 0.004 ms. On the other hand, the unitary processing rate starts around 0.04, decreasing continuously at the time in which the necessary resources are adequately allocated in memory.

1 https://github.com/mjdivan/streamCE

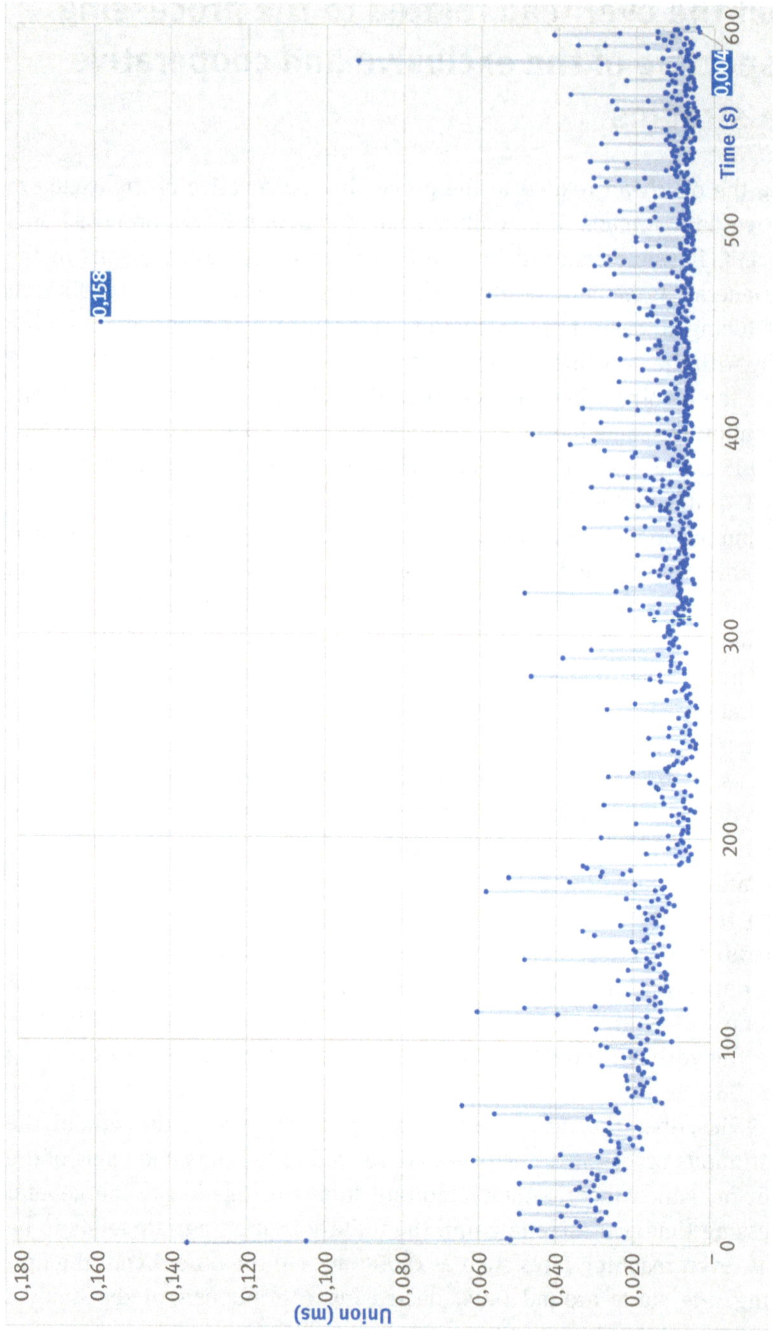


**Figure 6.9:** The unitary processing times of the union operation in the streamCE library.

Figure 6.9 expresses the elapsed time of the simulation on the axis of the abscissas in seconds, while the axis of the ordinate represents the unitary processing time for the union operation expressed in milliseconds. The peaks observed in the figure are related to the additional consumed time of the garbage collector. However, it is possible to appreciate through the concentration of points the general behavior and its tendency to progressively decrease the unitary rate up to reach values around 0.004 ms per operation.

Similarly, the projection operation was simulated throughout 10 min using a data stream named *streamA* that contained the numeric attributes "*a*," "*b*," "*c*," "*d*," "*e*," "*f*," "*g*," "*h*," and "*i*." The result consisted of a new cooperative data stream with the following attributes ordered as follows: "*c*," "*e*," "*i*," and "*h*."

Figure 6.10 represents the elapsed time of the simulation on the axis of the abscissas in seconds, while the axis of the ordinate represents the unitary processing time for the projection operation expressed in milliseconds. The peaks observed in the figure are related to the additional consumed time of the garbage collector.

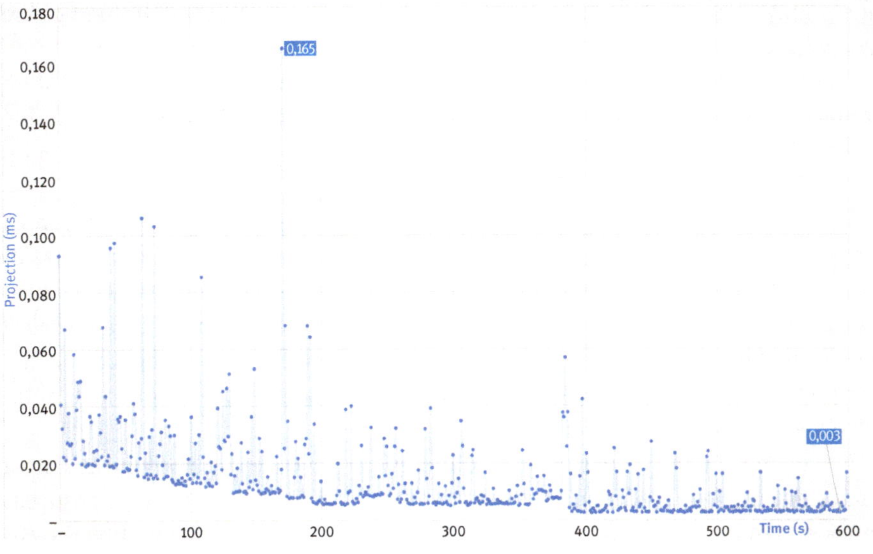


**Figure 6.10:** The unitary processing times of the projection operation in the streamCE library.

The individual processing rates started around 0.04 ms, progressively decreasing at the time in which all the necessary resources were allocated in memory. Near to the 10 min, the unitary processing rate reached values around 0.003 ms per operation.

The library is open to anyone who wants to use or study it. This could be extended incorporating new operations, specializing data streams based on different kinds of requirements, among other aspects. The idea is to provide a conceptual implementation of the perspective given for the cooperative and exclusive data streams in order to materialize the concepts. The obtained times are limited to the simulation and they are useful for describing a conceptual idea but not for any kind of statistical generalization.

## 6.9 Conclusions

In this chapter a discussion around the concept of data streams, the interpretation of its meaning in different contexts, and the representation of its content was outlined. It was possible to appreciate the absence of a consensus about the idea of the content; some authors refer the data streams such as an unbounded sequence of tuples, while other authors understood the data stream as an unbounded sequence of data, independently its internal structure. This amplitude of interpretations generates a gray zone in which it is a challenge to provide a complete definition of the concept itself.

The current approaches related to different ways to interpret the content were introduced, while the SMS was performed forward to analyze the tendencies in this subject. SMS was driven on the Scopus database for works published from 2016 up to now.

The data stream environment is mainly focused on to measure and evaluates different entities, events, or aspects that are necessary for making decisions. In this sense, data-driven decision-making has emerged as a real alternative for supporting the decision-making processes in all kinds of habitats in which a decision needs to be taken. For that reason, Section 6.4 introduced the importance of the measurement framework along with the measurement process and the data-driven decision-making.

Advancing on the importance of different perspectives to implement active monitoring, the centered and distributed processing strategies were introduced. They were synthetically described, schematized and compared, highlighting the environments in which each one could be applied. The impact of current technologies related to fog computing, the underlying idea associated with its data processing and closeness to users to provide answers were analyzed.

In this way, the definition and description of the exclusive and cooperative data streams were incorporated jointly with its formal definition. They were compared, describing the effects and limitations of using each one. This allows to model the content of each data stream, describing the relationship between the timestamp and the rest of the contained data, how the data structure could be defined and the restrictions about each associated values' domain.

Also, a notation for managing cooperative and exclusive data streams were proposed jointly with how the operations could be defined. The operations naturally provide how the data streams could be processed and transformed from one type to another. For example, the union could receive one positional data stream jointly with a temporal data stream, giving a temporal data stream as a result. The operator allows extending the way in which each data stream could be transformed following a given logic (e.g., union and projection).

The *streamCE* library was released under the terms of the Apache 2 General Agreement License, implementing the concepts here described as a proof of concept. In this way, anyone could use or extend its content, contrasting the ideas here described and even extending them to other data stream platforms.

Finally, a simulation on the projection and union operations was carried out on common hardware, using cooperative data streams and processing it over 10 min. The unitary processing rates were decreasing in each case at the time in which the necessary processing resources were allocated in memory, reaching unitary rates around 0.003 and 0.004 ms per operation for the projection and union, respectively.

As future work, application scenarios for cooperative and exclusive data streams will be analyzed, fostering its implementation on other data stream platforms (e.g., Apache Storm or Apache Spark).

# Bibliography

[1] Von Bertalanffy, L. (1972). The History and Status of General Systems Theory, Academy of Management Journal, 15(4), 407–426. https://doi.org/10.5465/255139.

[2] Bassett, D.S., Wymbs, N.F., Porter, M.A., Mucha, P.J., Carlson, J.M. and Grafton, S.T. (2011). Dynamic reconfiguration of human brain networks during learning, Proceedings of the National Academy of Sciences, 108(18), 7641–7646. https://doi.org/10.1073/pnas.1018985108.

[3] Almeida, P. and Chase-Dunn, C. (2018). Globalization and Social Movements, Annual Review of Sociology, 44(1), 189–211. https://doi.org/10.1146/annurev-soc-073117-041307.

[4] Dolan, J.G. and Fraenkel, L. (2017). Shared Decision-Making, Multi-Criteria Decision Analysis to Support Healthcare Decisions, Cham, Springer International Publishing, 199–215. https://doi.org/10.1007/978-3-319-47540-0_11.

[5] Schwarting, W., Alonso-Mora, J. and Rus, D. (2018). Planning and Decision-Making for Autonomous Vehicles, Annual Review of Control, Robotics, and Autonomous Systems, 1(1), 187–210. https://doi.org/10.1146/annurev-control-060117-105157.

[6] Ramírez-Gallego, S., Krawczyk, B., García, S., Woźniak, M. and Herrera, F. (2017). A survey on data preprocessing for data stream mining: Current status and future directions, Neurocomputing, 239, 39–57. https://doi.org/10.1016/j.neucom.2017.01.078.

[7] Safaei, A.A. (2017). Real-time processing of streaming big data, Real-Time Systems, 53(1), 1–44. https://doi.org/10.1007/s11241-016-9257-0.
[8] Khan, M.A. and Salah, K. (2018). IoT security: Review, blockchain solutions, and open challenges, Future Generation Computer Systems, 82, 395–411. https://doi.org/10.1016/j.future.2017.11.022.
[9] Ammar, M., Russello, G. and Crispo, B. (2018). Internet of Things: A survey on the security of IoT frameworks, Journal of Information Security and Applications. https://doi.org/10.1016/j.jisa.2017.11.002.
[10] Ni, J., Zhang, K., Lin, X. and Shen, X.S. (2018). Securing Fog Computing for Internet of Things Applications: Challenges and Solutions, IEEE Communications Surveys & Tutorials, 20(1), 601–628. https://doi.org/10.1109/COMST.2017.2762345.
[11] Ge, M., Bangui, H. and Buhnova, B. (2018). Big Data for Internet of Things: A Survey, Future Generation Computer Systems. https://doi.org/10.1016/j.future.2018.04.053.
[12] Reyna, A., Martín, C., Chen, J., Soler, E. and Díaz, M. (2018). On blockchain and its integration with IoT. Challenges and opportunities, Future Generation Computer Systems. https://doi.org/10.1016/j.future.2018.05.046.
[13] Jang, J., Jung, I. and Park, J.H. (2018). An effective handling of secure data stream in IoT, Applied Software Computer, 68, 811–820. https://doi.org/10.1016/j.asoc.2017.05.020.
[14] Nakazawa, J., Tokuda, H. and Yonezawa, T. (2015). Sensorizer: An Architecture for Regenerating Cyber Physical Data Streams from the Web. In Adjunct Proceedings of the 2015 ACM International Joint Conference on Pervasive and Ubiquitous Computing and Proceedings of the 2015 ACM International Symposium on Wearable Computers (pp. 1599–1606). New York, NY, USA: ACM. https://doi.org/10.1145/2800835.2801627
[15] Chen, G.J., Wiener, J.L., Iyer, S., Jaiswal, A., Lei, R., Simha, N. . . . Yilmaz, S. (2016). Realtime Data Processing at Facebook. In Proceedings of the 2016 International Conference on Management of Data (pp. 1087–1098). New York, NY, USA: ACM. https://doi.org/10.1145/2882903.2904441
[16] Rivetti, N., Busnel, Y. and Querzoni, L. (2016). Load-aware Shedding in Stream Processing Systems. In Proceedings of the 10th ACM International Conference on Distributed and Event-based Systems (pp. 61–68). New York, NY, USA: ACM. https://doi.org/10.1145/2933267.2933311
[17] Gulisano, V., Nikolakopoulos, Y., Cederman, D., Papatriantafilou, M. and Tsigas, P. (2017). Efficient Data Streaming Multiway Aggregation Through Concurrent Algorithmic Designs and New Abstract Data Types, ACM Transactions Parallel Computer, 4(2), 11, 1–11. 28. https://doi.org/10.1145/3131272.
[18] Chothia, Z., Liagouris, J., Dimitrova, D. and Roscoe, T. (2017). Online Reconstruction of Structural Information from Datacenter Logs. In Proceedings of the Twelfth European Conference on Computer Systems (pp. 344–358). New York, NY, USA: ACM. https://doi.org/10.1145/3064176.3064195
[19] Brandt, T. and Grawunder, M. (2018). GeoStreams: A Survey, ACM Computer Survey, 51(3), 44, 1–44. 37. https://doi.org/10.1145/3177848.
[20] Ivanov, T. and Taaffe, J. (2018). Exploratory Analysis of Spark Structured Streaming. In Companion of the 2018 ACM/SPEC International Conference on Performance Engineering (pp. 141–146). New York, NY, USA: ACM. https://doi.org/10.1145/3185768.3186360
[21] Ganibardi, A. and Ali, C.A. (2018). Weblog Data Structuration: A Stream-centric Approach for Improving Session Reconstruction Quality. In Proceedings of the 20th International Conference on Information Integration and Web-based Applications & Services (pp. 263–271). New York, NY, USA: ACM. https://doi.org/10.1145/3282373.3282379

[22] Chandy, K.M. and Bunn, J. (2019). Compositional Structures for Streaming Applications. In Proceedings of the 20th International Conference on Distributed Computing and Networking (pp. 352–361). New York, NY, USA: ACM. https://doi.org/10.1145/3288599.3288642
[23] Géraud, R., Lombard-Platet, M. and Naccache, D. (2019). Quotient Hash Tables: Efficiently Detecting Duplicates in Streaming Data. In Proceedings of the 34th ACM/SIGAPP Symposium on Applied Computing (pp. 582–589). New York, NY, USA: ACM. https://doi.org/10.1145/3297280.3297335
[24] Kitchenham, B.A., Budgen, D. and Pearl Brereton, O. (2011). Using mapping studies as the basis for further research – A participant-observer case study, Information and Software Technology, 53(6), 638–651. https://doi.org/10.1016/j.infsof.2010.12.011.
[25] Petersen, K., Vakkalanka, S. and Kuzniarz, L. (2015). Guidelines for conducting systematic mapping studies in software engineering: An update, Information and Software Technology, 64, 1–18. https://doi.org/10.1016/j.infsof.2015.03.007.
[26] Lughofer, E., Pratama, M. and Skrjanc, I. (2018). Incremental rule splitting in generalized evolving fuzzy systems for autonomous drift compensation, IEEE Transactions on Fuzzy Systems, 26(4), 1854–1865. https://doi.org/10.1109/TFUZZ.2017.2753727.
[27] Lughofer, E. (2017). On-line active learning: A new paradigm to improve practical useability of data stream modeling methods, Information Sciences, 415–416, 356–376. https://doi.org/10.1016/j.ins.2017.06.038.
[28] Georis-Creuseveau, J., Claramunt, C. and Gourmelon, F. (2017). A modelling framework for the study of Spatial Data Infrastructures applied to coastal management and planning, International Journal of Geographical Information Science, 31(1), 122–138. https://doi.org/10.1080/13658816.2016.1188929.
[29] Kamburugamuve, S., Wickramasinghe, P., Ekanayake, S. and Fox, G.C. (2018). Anatomy of machine learning algorithm implementations in MPI, Spark, and Flink, International Journal of High Performance Computing Applications, 32(1), 61–73. https://doi.org/10.1177/1094342017712976.
[30] Koek, P., Geuns, S.J., Hausmans, J.P.H.M., Corporaal, H. and Bekooij, M.J.G. (2016). CSDF$^a$: A model for exploiting the trade-off between data and pipeline parallelism. In Proceedings of the 19th International Workshop on Software and Compilers for Embedded Systems, SCOPES 2016 (pp. 30–39). https://doi.org/10.1145/2906363.2906364
[31] Dubrulle, P., Gaston, C., Kosmatov, N., Lapitre, A. and Louise, S. (2019). A data flow model with frequency arithmetic. Lecture Notes in Computer Science (including subseries Lecture Notes in Artificial Intelligence and Lecture Notes in Bioinformatics), Vol. 11424 LNCS. https://doi.org/10.1007/978-3-030-16722-6_22.
[32] Meinig, M., Tröger, P. and Meinel, C. (2019). Finding classification zone violations with anonymized message flow analysis. In ICISSP 2019 – Proceedings of the 5th International Conference on Information Systems Security and Privacy (pp. 284–292).
[33] Masulli, F. and Rovetta, S. (2019). The Challenges of Big Data and the Contribution of Fuzzy Logic. Lecture Notes in Computer Science (including subseries Lecture Notes in Artificial Intelligence and Lecture Notes in Bioinformatics), Vol. 11291 LNAI. https://doi.org/10.1007/978-3-030-12544-8_25.
[34] Abdullatif, A., Masulli, F. and Rovetta, S. (2018). Clustering of nonstationary data streams: A survey of fuzzy partitional methods, Wiley Interdisciplinary Reviews: Data Mining and Knowledge Discovery, 8(4). https://doi.org/10.1002/widm.1258.
[35] Abdullatif, A., Masulli, F. and Rovetta, S. (2017). Tracking Time Evolving Data Streams for Short-Term Traffic Forecasting, Data Science and Engineering, 2(3), 210–223. https://doi.org/10.1007/s41019-017-0048-y.

[36] Chadli, N., Kabbaj, M.I. and Bakkoury, Z. (2018). Detection of dataflow anomalies in business process an overview of modeling approaches. In ACM International Conference Proceeding Series. https://doi.org/10.1145/3289402.3289537

[37] Mackie, I. (2017). A geometry of interaction machine for gödel's system t. Lecture Notes in Computer Science (including subseries Lecture Notes in Artificial Intelligence and Lecture Notes in Bioinformatics), Vol. 10388 LNCS. https://doi.org/10.1007/978-3-662-55386-2_16.

[38] Godin, B. (2002). Outline for a history of science measurement, Science Technology and Human Values. https://doi.org/10.1177/016224390202700101.

[39] Tramontini, C.C. and Graziano, K.U. (2007). Hypothermia control in elderly surgical patients in the intraoperative period: evaluation of two nursing interventions, Revista Latino-Americana de Enfermagem, 15(4), 626–631. https://doi.org/10.1590/S0104-11692007000400016.

[40] Witkowski, K. (2017). Internet of Things, Big Data, Industry 4.0 – Innovative Solutions in Logistics and Supply Chains Management, Procedia Engineering, 182, 763–769. https://doi.org/10.1016/j.proeng.2017.03.197.

[41] Diván, M.J. (2017). Real-Time Measurement and Evaluation as System Reliability Driver, System Reliability Management, Vol. 4, Boca Raton, Taylor & Francis, a CRC title, part of the Taylor &: CRC Press, 161–188. https://doi.org/10.1201/9781351117661-11.

[42] Pan, T. and Zhu, Y. (2018). Getting Started with Arduino, Designing Embedded Systems with Arduino, Singapore, Springer Singapore, 3–16. https://doi.org/10.1007/978-981-10-4418-2_1.

[43] Glória, A., Cercas, F. and Souto, N. (2017). Design and implementation of an IoT gateway to create smart environments, Procedia Computer Science, 109, 568–575. https://doi.org/10.1016/j.procs.2017.05.343.

[44] Chakravarthy, S. and Jiang, Q. (2009). Stream Data Processing: A Quality of Service Perspective. Processing, Vol. 36, Boston, MA, Springer US. https://doi.org/10.1007/978-0-387-71003-7.

[45] Chaudhry, N. (2005). Stream Data Management, Chaudhry, N.A., Shaw, K. and Abdelguerfi, M., Eds., Database, Vol. 30, Boston, MA, Springer US. https://doi.org/10.1007/b106968.

[46] Oguz, D., Yin, S., Hameurlain, A., Ergenc, B. and Dikenelli, O. (2016). Adaptive Join Operator for Federated Queries over Linked Data Endpoints, Lecture Notes in Computer Science (including subseries Lecture Notes in Artificial Intelligence and Lecture Notes in Bioinformatics), 275–290. https://doi.org/10.1007/978-3-319-44039-2_19.

[47] Codd, E.F. (1983). A relational model of data for large shared data banks, Communications of the ACM, 26(1), 64–69. https://doi.org/10.1145/357980.358007.

[48] Pilman, M., Kaufmann, M., Köhl, F., Kossmann, D. and Profeta, D. (2016). ParTime: Parallel temporal aggregation. In Proceedings of the ACM SIGMOD International Conference on Management of Data. https://doi.org/10.1145/2882903.2903732

[49] Santos, M.Y. and Costa, C. (2016). Data models in NoSQL databases for big data contexts, Lecture Notes in Computer Science (including subseries Lecture Notes in Artificial Intelligence and Lecture Notes in Bioinformatics). https://doi.org/10.1007/978-3-319-40973-3_48.

[50] Zhao, G., Huang, W., Liang, S. and Tang, Y. (2013). Modeling MongoDB with relational model. In Proceedings – 4th International Conference on Emerging Intelligent Data and Web Technologies, EIDWT 2013. https://doi.org/10.1109/EIDWT.2013.25

Kuldeep Nagiya and Mangey Ram

# 7 Performance evaluation of switched telephone exchange network

**Abstract:** Public switched telephone network is the most useful telecommunication network. This model calculates the different reliability standard of a basic telephone exchange system. This model contains the subscriber loop systems. In this model, various subsystems are also included such as distributed cables (DC), feeder point (FP), main distribution frame (MDF), and distribution point (DP). In this system, DC are connected with FP, feeder cables are connected with MDF, and the drop wires are connected to wire pairs in DP. All the failure and service rates are general. The different reliability characteristics are found by using Laplace transformation and Markov process.

**Keywords:** Telephone exchange, system integration, fault-tolerant systems, switched network

## 7.1 Introduction

A telephone network is the most useful telecommunication network. The complexity of the telephone network is managed by using a hierarchical structure, worldwide standardization, and decentralization of administration, operation, and maintenance. Generally, four levels of cabling are used. The drop wires are connected to a distribution point (DP) at the subscriber end. The drop wires are connected to wire pairs in the distribution cables (DC) at the DP. In a feeder point (FP), various DC are terminated. The feeder cables are terminated on a main distribution frame (MDF) at the telephone exchange. Subscriber pairs and exchange pairs are interconnected at MDF by means of jumpers and MDF provides a flexible interconnection mechanism.

Many scholars have done a lot of research in this field. Flowers [1] studied on electronic telephone exchange and comparing the automatic telephone exchange and exchange using cheaper and more reliabile electromechanical switches. He also discussed about the possibility of using time division multiplex and multiple frequency division type of connector switches for speech and line signal transmission. Palmer [2] discussed the principles for maintaining the automatic telephone exchange plant and analyzed that the maintenance of equipment that are used in

**Kuldeep Nagiya, Mangey Ram,** Department of Mathematics, Graphic Era University, Dehradun, Uttarakhand, India

https://doi.org/10.1515/9783110619058-007

automatic telephone exchange was very easy than the maintenance of equipment that are used in earlier one because in earlier the error finding is very much difficult. He suggested that the preventive maintenance was required for routine checkup of the condition of the plant. Depp and Townsend [3] studied on electronic branch telephone exchange switching system. They described the overall system design and operation. This study tells the superiority of the built-in speed of electronic devices. Warman and Bear [4] analyzed the trunking and traffic aspects of a telephone exchange system, which was divided into parts. They described a study for the design of a telephone exchange system in which a network is divided into different parts. They also determined that which part is best for each call. They also described the working of reed relay exchange system that was controlled electronically. Strandberg and Ericsson [5] analyzed the reliability prediction in telephone system engineering. They discussed some aspects of the definition and application of new concepts and the classification and choice of appropriate measures. They measured the maintainability, traffic ability, availability, and system effectiveness, quality of measures as well as reliability of the system. Malec [6] analyzed the reliability optimization in design of a telephone switching system and described that the optimization techniques are used in both allocation and modeling. He also discussed about the objectives of reliability and different methods of reliability allocation. Baron et al. [7] studied about the behavior of telephone relays and connectors in various antipathetic environments. They conducted three types of studies. A study on atmospheric analysis in telephone offices, surface analysis of contacts from telephone exchanges, and a study on laboratory simulation tests. Tortorella [8] discussed the cutoff calls and the telephone equipment reliability, and described a mathematical model for the rate of cutoff calls caused by failure and malfunctions in telephone equipment. This model compared the rate of cutoff calls produced by failures in the equipment and its subsystems to the failure modes in the equipment, their severity and frequency of occurrence, and the call holding time distribution. The author used a model as a $c$-server queuing system. Lelievre and Goarin [9] discussed the consequence conditions on the accuracy in working of electronic components in telephone exchanges. They collected numerous data and used sophisticated data processing in this study. They conducted a physical analysis for the components failure and also studied about factors affecting the reliability of the components. Kolte [10] described the cooling system BPA-105 for small telephone exchanges and described that the cooling system must meet the following requirement such as high reliability, cooling reserves during mains failure and minimum maintenance. The different system units such as cooling, pump, control, supervision, and ventilation and stated that the cooling system has the following features easy installation, small risk of condensation, good environment and minimum maintenance. Kanoun et al. [11] developed software for reliability analysis and prediction to the TROPICO-R switching system. They presented a method that allows living reliability growth models. They described that the hyper exponential mode is allowed forecasting of the software residual failure rate. This method is applied to the Brazilian electronic switching system TROPICO-R. Fagerstrom and Healy [12] discussed the reliability of local exchange carrier telephone network. They described that the reliability of local exchange carrier (LEC) networks was obtained from Bell core's Outage Performance Monitoring (OPM) processes. Kuhn [13] described the roots of failure in the public switched telephone network. The failures in the public switched telephone network were due to the human intervention. Hellstrom et al. [14] described the method of cooling for telephone exchanges using borehole heat exchanges in various atmospheres. They described that there is no need of cooling machine, and the consumption of electric power is very low. The main advantages of this system are very high air temperatures, excellent reliability, and very low cost of maintenance. They developed a system that contains factory-assembled system control unit, atmosphere cooling unit, outdoor recooling unit, ground cooling unit using borehole heat exchange. Mirjalily et al. [15, 16] analyzed the centers of telephone exchange over metro Ethernet networks by using DiffServ-MPLS. They studied the performance of connecting telephone exchange centers over a metro Ethernet network using MPLS and different server quality of service model and implemented some simulation techniques to evaluate the performance in terms of delay, jitter, and loss. They also discussed the performance evaluation of different quality of service models for connecting telephone exchange centers over metro Ethernet networks.

## 7.2 System description, assumptions, and notations

Initially, all the components are in working conditions. If the FP fails, then the system goes into a degraded or partial failed state. If the DP fails, then the system will goes into a partially failed stage. The system is completely failed due to failure of MDF. The system has also completely failed due to the failure of the power station and DC respectively. If either the DP is failed after the failure of FP or the FP is failed after the failure of DP then the system is also completely failed. The flow of failure can be seen in transition state diagram in Figure 7.1.

In this model, the following assumptions are taken:

(i) All the subsystems are in good working conditions.
(ii) Three types of states are considered: good, partially, and completely failed states.
(iii) Constant failure and repair rates.
(iv) Sufficient repair facilities are available.
(v) After repairing, the model goes in working conditions.

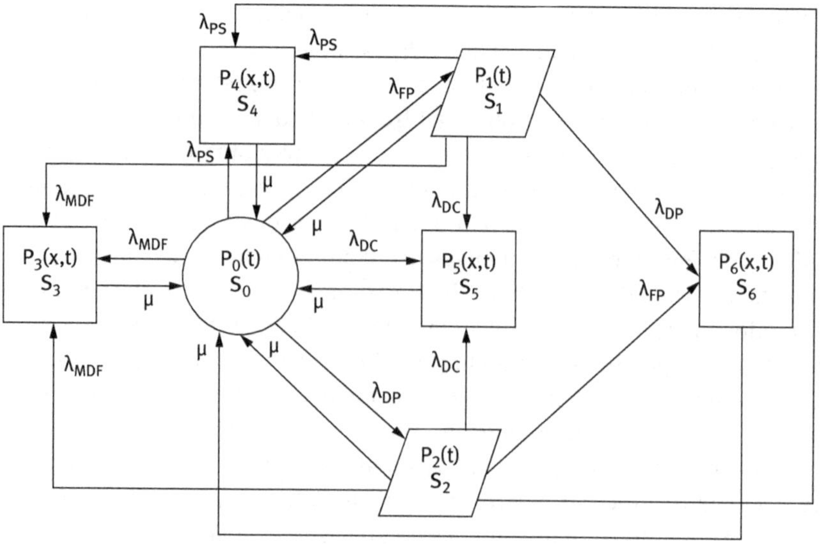


**Figure 7.1:** Transition state diagram.

We use the following notations in this model:

| | |
|---|---|
| $t$ | Time unit |
| $s$ | Variable used in Laplace transforms |
| $P_0\ (t)$ | Probability at time $t$ in initial state |
| $P_i\ (t)$ | Probability at time $t$ in $i$th state; $i = 1, 2$ |
| $P_j\ (x,t)$ | P.d.f. of the system is in $j$th state where $j = 3, 4, 5, 6$ |
| $\lambda_{MDF}$ | Main distribution frame failure rate |
| $\lambda_{PS}$ | Power supply failure rate |
| $\lambda_{DP}$ | Distributed point failure rate |
| $\lambda_{FP}$ | Feeder point failure rate |
| $\lambda_{DC}$ | Distributed cable failure rate |
| $\mu$ | Repairing rate |

The descriptions of various states involved in the designed system have been shown in the following Table 7.1.

**Table 7.1:** State descriptions.

| State | Descriptions of different state |
|---|---|
| $S_0$ | This is the starting stage of the model. All subsystems are in perfectly good working conditions. |
| $S_1$ | In $S_1$, subsystem feeder point is failed and other subsystems are in working conditions, therefore this state is partially failed. |

**Table 7.1** (continued)

| State | Descriptions of different state |
|---|---|
| $S_2$ | In $S_2$, subsystem distributed point is failed and other subsystems are in working conditions, therefore this state is also partially failed. |
| $S_3$ | In $S_3$, subsystem main distribution frame is failed and other subsystems are in working conditions, therefore the state $S_3$ is completely failed state. |
| $S_4$ | In $S_4$, subsystem power supply is failed and other subsystems are in working conditions, therefore this state is also completely failed state. |
| $S_5$ | In $S_5$, subsystem distributed cable is failed and other subsystems are in working conditions, therefore this state is also completely failed state. |
| $S_6$ | In $S_6$, subsystem feeder point and distributed point are failed and other subsystems are in working conditions, therefore this state is also completely failed state. |

## 7.3 Formation and solution of the differential equations

On the basis of following transition state diagram, we developed the following set of differential equations:

$$\left[\frac{\partial}{\partial t}+\lambda_{\mathrm{PS}}+\lambda_{\mathrm{FP}}+\lambda_{\mathrm{DC}}+\lambda_{\mathrm{DP}}+\lambda_{\mathrm{MDF}}\right]P_0(t)=\mu\{P_1(t)+P_2(t)\}+\sum_j\int_0^\infty P_j(x,t)\mu dx;\ j=3\text{ to }6 \tag{7.1}$$

$$\left[\frac{\partial}{\partial t}+\lambda_{\mathrm{PS}}+\lambda_{\mathrm{DC}}+\lambda_{\mathrm{DP}}+\lambda_{\mathrm{MDF}}+\mu\right]P_1(t)=\lambda_{\mathrm{FP}}P_0(t) \tag{7.2}$$

$$\left[\frac{\partial}{\partial t}+\lambda_{\mathrm{PS}}+\lambda_{\mathrm{DC}}+\lambda_{\mathrm{DP}}+\lambda_{\mathrm{MDF}}+\mu\right]P_2(t)=\lambda_{\mathrm{DP}}P_0(t) \tag{7.3}$$

$$\left[\frac{\partial}{\partial t}+\frac{\partial}{\partial x}+\mu\right]P_3(x,t)=0 \tag{7.4}$$

$$\left[\frac{\partial}{\partial t}+\frac{\partial}{\partial x}+\mu\right]P_4(x,t)=0 \tag{7.5}$$

$$\left[\frac{\partial}{\partial t}+\frac{\partial}{\partial x}+\mu\right]P_5(x,t)=0 \tag{7.6}$$

$$\left[\frac{\partial}{\partial t}+\frac{\partial}{\partial x}+\mu\right]P_6(x,t)=0 \tag{7.7}$$

We are using the different boundary conditions:

$$P_3(0,t) = \lambda_{MDF}\{P_0(t) + P_1(t) + P_2(t)\} \tag{7.8}$$

$$P_4(0,t) = \lambda_{PS}\{P_0(t) + P_1(t) + P_2(t)\} \tag{7.9}$$

$$P_5(0,t) = \lambda_{DC}\{P_0(t) + P_1(t) + P_2(t)\} \tag{7.10}$$

$$P_6(0,t) = \lambda_{DP}P_1(t) + \lambda_{FP}P_2(t) \tag{7.11}$$

We know that at $t = 0$, $P_0(t) = 1$ and other probabilities are zero.

Now, we take the Laplace transformation of eqs. (7.1) to (7.11):

$$[s + \lambda_{PS} + \lambda_{FP} + \lambda_{DC} + \lambda_{DP} + \lambda_{MDF}]\overline{P}_0(s) = 1 + \mu\{\overline{P}_1(s) + \overline{P}_2(s)\} + \sum_j \int_0^\infty \overline{P}_j(x,s)\mu dx;\ j = 3 \text{ to } 6 \tag{7.12}$$

$$[s + \lambda_{PS} + \lambda_{DC} + \lambda_{DP} + \lambda_{MDF} + \mu]\overline{P}_1(s) = \lambda_{FP}\overline{P}_0(s) \tag{7.13}$$

$$[s + \lambda_{PS} + \lambda_{DC} + \lambda_{DP} + \lambda_{MDF} + \mu]\overline{P}_2(s) = \lambda_{DP}\overline{P}_0(s) \tag{7.14}$$

$$\left[s + \frac{\partial}{\partial x} + \mu\right]\overline{P}_3(x,s) = 0 \tag{7.15}$$

$$\left[s + \frac{\partial}{\partial x} + \mu\right]\overline{P}_4(x,s) = 0 \tag{7.16}$$

$$\left[s + \frac{\partial}{\partial x} + \mu\right]\overline{P}_5(x,s) = 0 \tag{7.17}$$

$$\left[s + \frac{\partial}{\partial x} + \mu\right]\overline{P}_6(x,s) = 0 \tag{7.18}$$

From boundary conditions,

$$\overline{P}_3(0,s) = \lambda_{MDF}\{\overline{P}_0(s) + \overline{P}_1(s) + \overline{P}_2(s)\} \tag{7.19}$$

$$\overline{P}_4(0,s) = \lambda_{PS}\{\overline{P}_0(s) + \overline{P}_1(s) + \overline{P_2}(s)\} \tag{7.20}$$

$$\overline{P}_5(0,s) = \lambda_{DC}\{\overline{P}_0(s) + \overline{P}_1(s) + \overline{P_2}(s)\} \tag{7.21}$$

$$\overline{P}_6(0,s) = \lambda_{DP}\overline{P}_1(s) + \lambda_{FP}\overline{P}_2(s) \tag{7.22}$$

From eqs. (7.15) to (7.18), we get,

$$\overline{P}_j(x,s) = \overline{P}_j(0,s)\exp\{-sx - \int_0^x \mu dx\};\ j = 3 \text{ to } 6 \tag{7.23}$$

Solution for the differential equations is given as:

$$\overline{P}_0(s) = \frac{1}{c - \mu c_1 - \mu c_2 - \{(\lambda_{MDF} + \lambda_{PS} + \lambda_{DC})(1 + c_1 + c_2) + 2\lambda_{DP} c_1\}\overline{S}(s)} \quad (7.24)$$

$$\overline{P}_1(s) = c_1 \overline{P}_0(s) \quad (7.25)$$

$$\overline{P}_2(s) = c_2 \overline{P}_0(s) \quad (7.26)$$

$$\overline{P}_3(s) = \lambda_{MDF}(1 + c_1 + c_2)\left(\frac{1 - \overline{S}(s)}{s}\right)\overline{P}_0(s) \quad (7.27)$$

$$\overline{P}_4(s) = \lambda_{PS}(1 + c_1 + c_2)\left(\frac{1 - \overline{S}(s)}{s}\right)\overline{P}_0(s) \quad (7.28)$$

$$\overline{P}_5(s) = \lambda_{DC}(1 + c_1 + c_2)\left(\frac{1 - \overline{S}(s)}{s}\right)\overline{P}_0(s) \quad (7.29)$$

$$\overline{P}_6(s) = 2\lambda_{DP} c_1\left(\frac{1 - \overline{S}(s)}{s}\right)\overline{P}_0(s) \quad (7.30)$$

where

$$c = s + \lambda_{PS} + \lambda_{FP} + \lambda_{DC} + \lambda_{DP} + \lambda_{MDF} \quad (7.31)$$

$$c_1 = \frac{\lambda_{FP}}{s + \lambda_{PS} + \lambda_{DP} + \lambda_{DC} + \lambda_{MDF} + \mu} \quad (7.32)$$

$$c_2 = \frac{\lambda_{DP}}{s + \lambda_{PS} + \lambda_{DP} + \lambda_{DC} + \lambda_{MDF} + \mu} \quad (7.33)$$

The up and down state probabilities at any time '*t*' are given later:

$$\overline{P}_{up}(s) = \overline{P}_0(s) + \overline{P}_1(s) + \overline{P}_2(s) = (1 + c_1 + c_2)\overline{P}_0(s) \quad (7.34)$$

$$\overline{P}_{down}(s) = \overline{P}_3(s) + \overline{P}_4(s) + \overline{P}_5(s) + \overline{P}_6(s)$$

$$= \{(\lambda_{MDF} + \lambda_{PS} + \lambda_{DC})(1 + c_1 + c_2) + 2\lambda_{DP} c_1\}\left(\frac{1 - \overline{S}(s)}{s}\right)\overline{P}_0(s) \quad (7.35)$$

## 7.4 Particular cases

In this section, the various reliability characteristics such as availability, reliability, MTTF, expected profit, as well as sensitivity analysis of the system are analyzed.

### 7.4.1 Availability

Availability is the probability of an operable system at a given specified time, that is, the amount of time a system is actually operating as the percentage of the total time, and it should be operated. For finding the availability of the system, we assume the different failure rates as $\lambda_{MDF} = 0.001$, $\lambda_{DP} = 0.003$, $\lambda_{FP} = 0.003$, $\lambda_{DC} = 0.004$, $\lambda_{PS} = 0.002$, and $\mu = 1$ in eq. (7.34) and after replacing these values, inverse Laplace transformation is calculated. We find the following availability of the system:

$$P_{up}(t) = 0.9930311886 + 1.987166831e^{-1.013t} - 1.98019802e^{-1.01t} \tag{7.36}$$

Increasing the time from $t = 0$ to $t = 15$ by unit increment in $P_{up}(t)$, we obtained the different values of availability that are shown in Table 7.2. With the help of these values a graph that shows the trend of availability of the system is illustrated in Figure 7.2.

**Table 7.2:** Availability as function of time.

| Time ($t$) | Availability $P_{up}(t)$ |
|---|---|
| 0 | 1.000000 |
| 1 | 0.993401 |
| 2 | 0.992379 |
| 3 | 0.992508 |
| 4 | 0.992737 |
| 5 | 0.992886 |
| 6 | 0.992965 |
| 7 | 0.993002 |
| 8 | 0.993019 |
| 9 | 0.993026 |
| 10 | 0.993029 |
| 11 | 0.993030 |
| 12 | 0.993030 |
| 13 | 0.993031 |
| 14 | 0.993031 |
| 15 | 0.993031 |

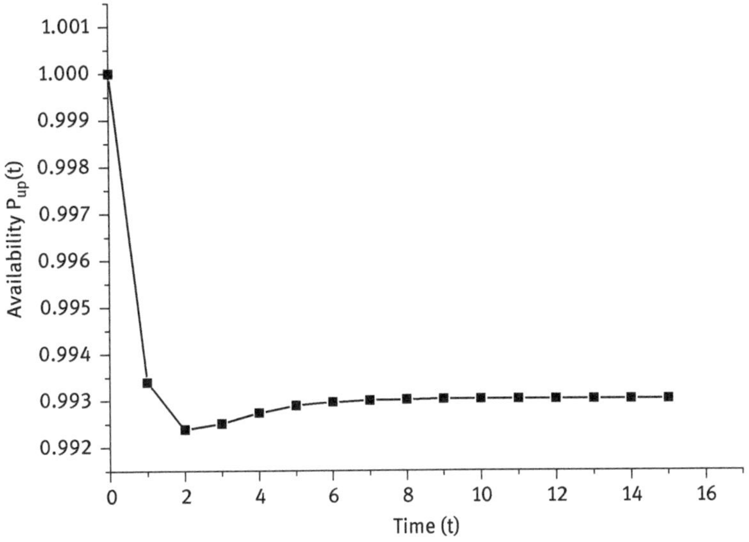


**Figure 7.2:** Availability as function of time.

## 7.4.2 Reliability

For finding the reliability of the system, we assume repair rate $\mu = 0$ in eq. (7.34). After that we will take the inverse Laplace transformation. The reliability of the system is given as follows:

$$R(t) = \frac{(\lambda_{DP} + \lambda_{FP})e^{-(\lambda_{MDF} + \lambda_{PS} + \lambda_{DP} + \lambda_{DC})t} - \lambda_{DP}e^{-(\lambda_{DC} + \lambda_{PS} + \lambda_{FP} + \lambda_{DP} + \lambda_{MDF})t}}{\lambda_{FP}} \tag{7.37}$$

Increasing the time from $t = 0$ to $t = 15$ by unit increment in $R(t)$, we obtained the different values of availability that are shown in Table 7.3. With the help of these values a graph that shows the trend of availability of the system is illustrated in Figure 7.3.

**Table 7.3:** Reliability as function of time.

| Time | Reliability |
|---|---|
| 0 | 1.000000 |
| 1 | 0.993016 |
| 2 | 0.986062 |
| 3 | 0.979140 |

**Table 7.3** (continued)

| Time | Reliability |
|---|---|
| 4 | 0.972250 |
| 5 | 0.965391 |
| 6 | 0.958565 |
| 7 | 0.951770 |
| 8 | 0.945007 |
| 9 | 0.938277 |
| 10 | 0.931579 |
| 11 | 0.924914 |
| 12 | 0.918281 |
| 13 | 0.911682 |
| 14 | 0.905115 |
| 15 | 0.898581 |

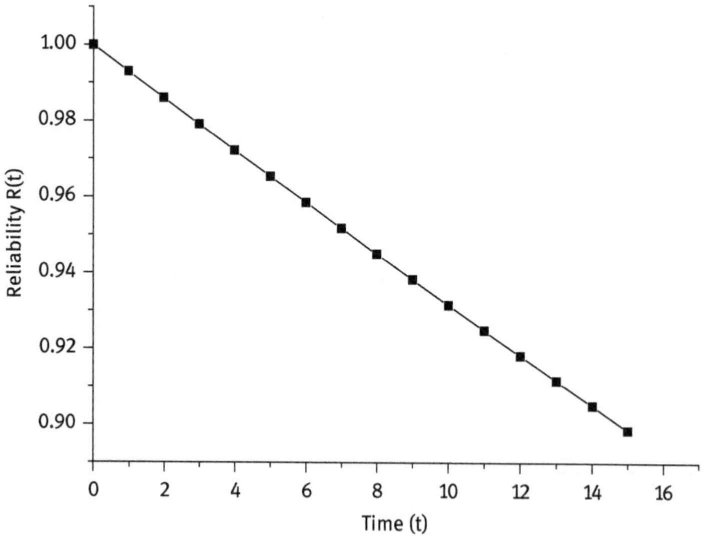


**Figure 7.3:** Reliability as function of time.

### 7.4.3 Mean time to failure (MTTF)

Mean time to failure (MTTF) is the length of time a device is expected to last in operation. MTTF represents how long a device can reasonably be expected to perform in the field based on specific testing.

To find the *MTTF* by letting the repair rate $\mu = 0$ and taking the Laplace variable's' tends to zero ($s \to 0$) in eq. (34):

$$\mathrm{MTTF} = \frac{\lambda_{PS} + \lambda_{DC} + \lambda_{MDF} + \lambda_{FP} + 2\lambda_{DP}}{(\lambda_{PS} + \lambda_{DP} + \lambda_{DC} + \lambda_{MDF})(\lambda_{PS} + \lambda_{DP} + \lambda_{DC} + \lambda_{MDF} + \lambda_{FP})} \tag{7.38}$$

Assuming $\lambda_{MDF} = 0.001$, $\lambda_{PS} = 0.002$, $\lambda_{DP} = 0.003$, $\lambda_{FP} = 0.003$, and $\lambda_{DC} = 0.004$ and varying $\lambda_{MDF}$, $\lambda_{PS}$, $\lambda_{DP}$, $\lambda_{FP}$, and $\lambda_{DC}$, respectively, as 0.1, 0.2, 0.3, 0.4, 0.5, 0.6, 0.7, 0.8, 0.9 in eq. (7.38), the variation of MTTF with respect to failure rates as shown in Table 7.4 and graphically in Figure 7.4.

**Table 7.4:** MTTF as function of different failure rates.

| Variation in $\lambda_{MDF}$, $\lambda_{PS}$, $\lambda_{DP}$, $\lambda_{FP}$, and $\lambda_{DC}$ | MTTF | | | | |
|---|---|---|---|---|---|
| | $\lambda_{MDF}$ | $\lambda_{PS}$ | $\lambda_{DP}$ | $\lambda_{FP}$ | $\lambda_{DC}$ |
| 0.1 | 9.420052 | 9.509509 | 17.841971 | 102.727273 | 9.693613 |
| 0.2 | 4.852397 | 4.876048 | 9.431792 | 101.428571 | 4.924049 |
| 0.3 | 3.267364 | 3.278072 | 6.409583 | 100.967742 | 3.299702 |
| 0.4 | 2.462791 | 2.468871 | 4.854078 | 100.731707 | 2.481121 |
| 0.5 | 1.976148 | 1.980061 | 3.906099 | 100.588235 | 1.987933 |
| 0.6 | 1.650085 | 1.652812 | 3.267885 | 100.491803 | 1.658293 |
| 0.7 | 1.416380 | 1.418389 | 2.808933 | 100.422535 | 1.422424 |
| 0.8 | 1.240661 | 1.242202 | 2.463016 | 100.370370 | 1.245295 |
| 0.9 | 1.103729 | 1.104948 | 2.192956 | 100.329670 | 1.107395 |

### 7.4.4 Expected profit

We can find the expected profit [17] in the interval (0, *t*] by using the equation,

$$E_P(t) = K_1 \int_0^t P_{up}(t)dt - tK_2 \tag{7.39}$$

In this equation, $K_1$ and $K_2$ represent the revenue and service cost per unit time, respectively.

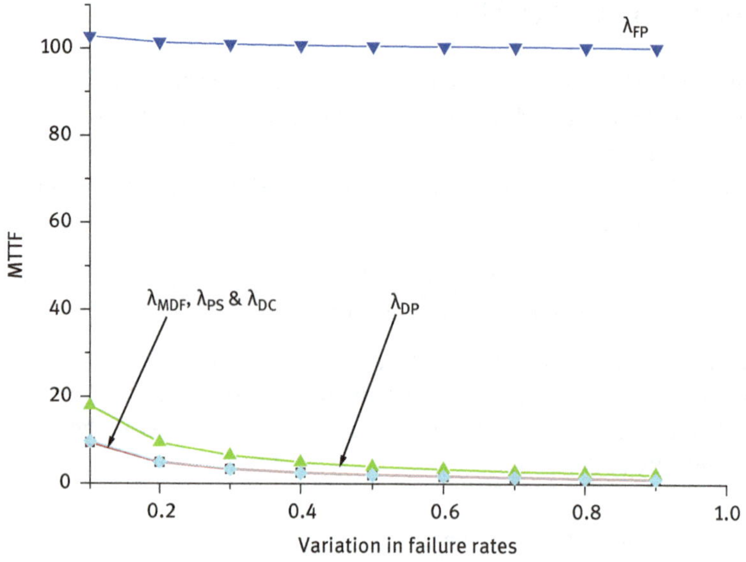


**Figure 7.4:** Graph of MTTF as function of different failure rates.

Putting the value of $P_{\mathrm{up}}(t)$ in eq. (7.39) and integrate within the limit, we obtained the expected:

$$E_p(t) = 0.9930311886t - 1.961665184e^{-1.013t} + 1.960592099e^{-1.01t} + 0.001073084603 - tK_2 \tag{7.40}$$

Putting $K_1 = 1$ and $K_2 = 0.01, 0.02, 0.04, 0.06, 0.08, 0.1, 0.2, 0.4$, respectively, in eq. (7.40), we get Table 7.5 and Figure 7.5, respectively.

**Table 7.5:** Values of expected profit at various service rates.

| Time ($t$) | Expected Profit $E_p(t)$ | | | | | | | |
|---|---|---|---|---|---|---|---|---|
| | $K_2 = 0.01$ | $K_2 = 0.02$ | $K_2 = 0.04$ | $K_2 = 0.06$ | $K_2 = 0.08$ | $K_2 = 0.1$ | $K_2 = 0.2$ | $K_2 = 0.4$ |
| 0 | 0.000000 | 0.000000 | 0.000000 | 0.000000 | 0.000000 | 0.000000 | 0.000000 | 0.000000 |
| 1 | 0.985854 | 0.975854 | 0.955854 | 0.935854 | 0.915854 | 0.895854 | 0.795854 | 0.595854 |
| 2 | 1.968549 | 1.948549 | 1.908549 | 1.868549 | 1.828549 | 1.788549 | 1.588549 | 1.188549 |
| 3 | 2.950964 | 2.920964 | 2.860964 | 2.800964 | 2.740964 | 2.680964 | 2.380964 | 1.780964 |
| 4 | 3.933591 | 3.893591 | 3.813591 | 3.733591 | 3.653591 | 3.573591 | 3.173591 | 2.373591 |
| 5 | 4.916409 | 4.866409 | 4.766409 | 4.666409 | 4.566409 | 4.466409 | 3.966409 | 2.966409 |

**Table 7.5** (continued)

| Time ($t$) | Expected Profit $E_p(t)$ | | | | | | | |
|---|---|---|---|---|---|---|---|---|
| | $K_2 = 0.01$ | $K_2 = 0.02$ | $K_2 = 0.04$ | $K_2 = 0.06$ | $K_2 = 0.08$ | $K_2 = 0.1$ | $K_2 = 0.2$ | $K_2 = 0.4$ |
| 6 | 5.899339 | 5.839339 | 5.719339 | 5.599339 | 5.479339 | 5.359339 | 4.759339 | 3.559339 |
| 7 | 6.882325 | 6.812325 | 6.672325 | 6.532325 | 6.392325 | 6.252325 | 5.552325 | 4.152325 |
| 8 | 7.865336 | 7.785336 | 7.625336 | 7.465336 | 7.305336 | 7.145336 | 6.345336 | 4.745336 |
| 9 | 8.848359 | 8.758359 | 8.578359 | 8.398359 | 8.218359 | 8.038359 | 7.138359 | 5.338359 |

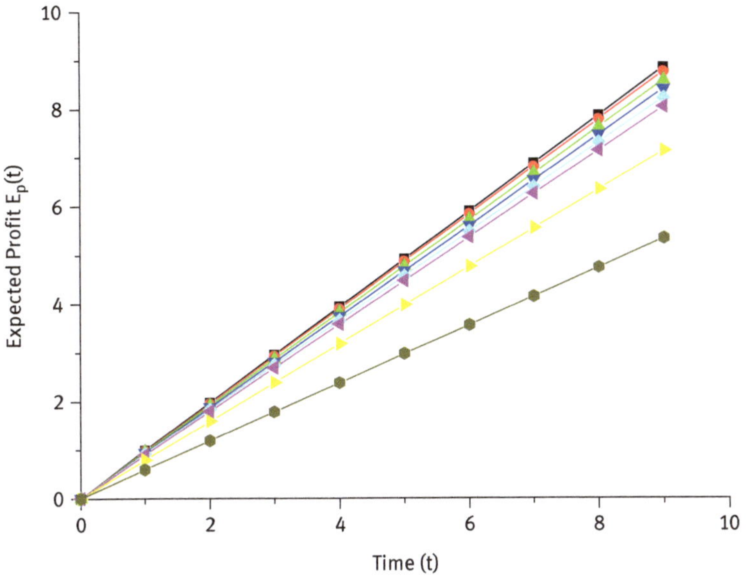


**Figure 7.5:** Graph of expected profit at various service rates.

## 7.4.5 Post optimality analysis (sensitivity analysis)

With the help of sensitivity analysis we can find out how sensitive an output is to any change in an input while keeping others input constant. Sensitivity analysis of a factor can be found by partial derivative of the function with respect to that factor. In the analyses, different failure rates represent the following factors [18].

**(i) Sensitivity of reliability**

To find the sensitivity analysis of reliability, we differentiate the equation of reliability partially with respect to the failure rate $\lambda_{MDF}$, $\lambda_{PS}$, $\lambda_{DP}$, $\lambda_{FP}$, and $\lambda_{DC}$, respectively, by using the values of $\lambda_{MDF} = 0.001$, $\lambda_{PS} = 0.002$, $\lambda_{DP} = 0.003$, $\lambda_{FP} = 0.003$, and $\lambda_{DC} = 0.004$. We find $\frac{\partial R(t)}{\partial \lambda_{MDF}}$, $\frac{\partial R(t)}{\partial \lambda_{PS}}$, $\frac{\partial R(t)}{\partial \lambda_{DP}}$, $\frac{\partial R(t)}{\partial \lambda_{FP}}$, and $\frac{\partial R(t)}{\partial \lambda_{DC}}$ and putting the value of $t = 0$ to $t = 10$, we observed the following values and graph in Table 7.6 and Figure 7.6, respectively.

**Table 7.6:** Sensitivity of reliability as a function of time.

| Time | $\frac{\partial R(t)}{\partial \lambda_{MDF}}$ | $\frac{\partial R(t)}{\partial \lambda_{PS}}$ | $\frac{\partial R(t)}{\partial \lambda_{DP}}$ | $\frac{\partial R(t)}{\partial \lambda_{FP}}$ | $\frac{\partial R(t)}{\partial \lambda_{DC}}$ |
|---|---|---|---|---|---|
| 0 | 0 | 0 | 0 | 0 | 0 |
| 1 | −0.993015 | −0.993015 | −0.004449 | −0.001482 | −0.993015 |
| 2 | −1.972124 | −1.972124 | −0.017596 | −0.005858 | −1.972124 |
| 3 | −2.937421 | −2.937421 | −0.039146 | −0.013023 | −2.937421 |
| 4 | −3.889000 | −3.889000 | −0.068809 | −0.022875 | −3.889000 |
| 5 | −4.826957 | −4.826957 | −0.106303 | −0.035316 | −4.826957 |
| 6 | −5.751388 | −5.751388 | −0.151352 | −0.050249 | −5.751388 |
| 7 | −6.662389 | −6.662389 | −0.203686 | −0.067579 | −6.662389 |
| 8 | −7.560059 | −7.560059 | −0.263043 | −0.087214 | −7.560059 |
| 9 | −8.444494 | −8.444494 | −0.329164 | −0.109064 | −8.444494 |
| 10 | −9.315794 | −9.315794 | −0.401798 | −0.133041 | −9.315794 |

**(ii) Sensitivity of MTTF**

To find the sensitivity analysis of MTTF, we differentiate the equation of MTTF partially with respect to the failure rate $\lambda_{MDF}$, $\lambda_{PS}$, $\lambda_{DP}$, $\lambda_{FP}$, and $\lambda_{DC}$, respectively, by using the values of $\lambda_{MDF} = 0.001$, $\lambda_{PS} = 0.002$, $\lambda_{DP} = 0.003$, $\lambda_{FP} = 0.003$, and $\lambda_{DC} = 0.004$. We find $\frac{\partial \mathrm{MTTF}}{\partial \lambda_{MDF}}$, $\frac{\partial \mathrm{MTTF}}{\partial \lambda_{PS}}$, $\frac{\partial \mathrm{MTTF}}{\partial \lambda_{DP}}$, $\frac{\partial \mathrm{MTTF}}{\partial \lambda_{FP}}$, and $\frac{\partial \mathrm{MTTF}}{\partial \lambda_{DC}}$. Now apply the variations in failure rates respectively as 0.1, 0.2, 0.3, 0.4, 0.5, 0.6, 0.7, 0.8, and 0.9 sensitivity of MTTF; we find the following values and observed a graph in Table 7.7 and Figure 7.7.

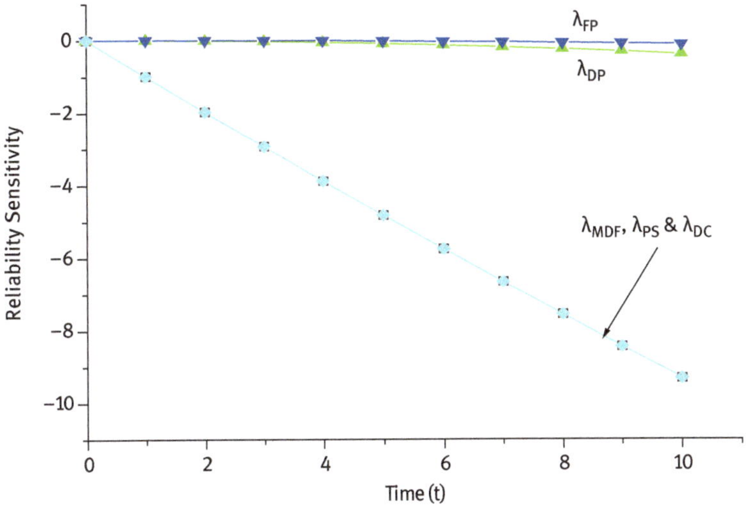


**Figure 7.6:** Graph of sensitivity of reliability as a function of time.

**Table 7.7:** Sensitivity of MTTF as a function of failure rates.

| Variation in $\lambda_{MDF}$, $\lambda_{PS}$, $\lambda_{DP}$, $\lambda_{FP}$, and $\lambda_{DC}$ | $\frac{\partial MTTF}{\partial \lambda_{MDF}}$ | $\frac{\partial MTTF}{\partial \lambda_{PS}}$ | $\frac{\partial MTTF}{\partial \lambda_{DP}}$ | $\frac{\partial MTTF}{\partial \lambda_{FP}}$ | $\frac{\partial MTTF}{\partial \lambda_{DC}}$ |
|---|---|---|---|---|---|
| 0.1 | -88.616611 | -90.305521 | -159.023597 | -24.793388 | -93.831289 |
| 0.2 | -23.536586 | -23.766500 | -44.468766 | -6.802721 | -24.236547 |
| 0.3 | -10.673729 | -10.743797 | -20.539168 | -3.121748 | -10.886018 |
| 0.4 | -6.064706 | -6.094683 | -11.780318 | -1.784652 | -6.155306 |
| 0.5 | -3.904896 | -3.920373 | -7.628505 | -1.153402 | -3.951605 |
| 0.6 | -2.722652 | -2.731658 | -5.339392 | -0.806235 | -2.749806 |
| 0.7 | -2.006062 | -2.011756 | -3.944972 | -0.595120 | -2.023218 |
| 0.8 | -1.539197 | -1.543023 | -3.033178 | -0.457247 | -1.550719 |
| 0.9 | -1.228191 | -1.220884 | -2.404498 | -0.362275 | -1.226298 |

## 7.5 Results discussion and conclusions

With the help of mathematical modeling, Laplace transformation, and supplementary variable technique, we obtain the different reliability characteristics of telephone exchange system. This model is useful for finding the different reliability characteristics

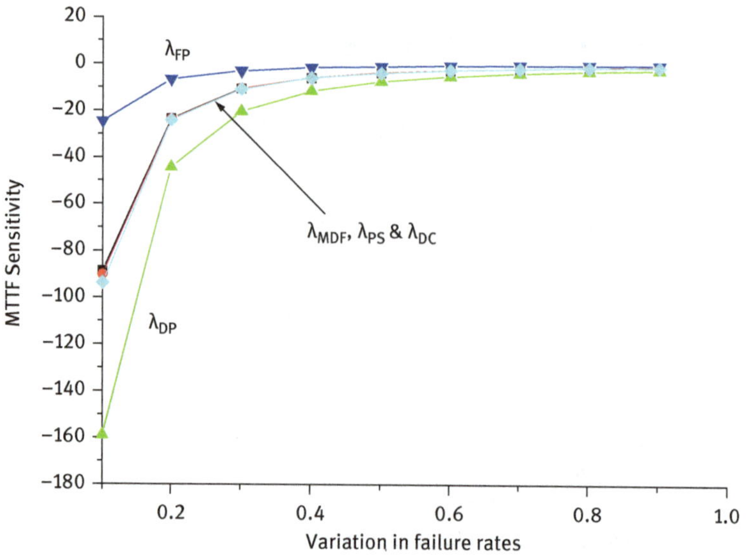


**Figure 7.7:** Graph of sensitivity of MTTF as a function of failure rates.

by assuming the various values of different failures. This model evaluates the availability, reliability, MTTF, expected profit, and post optimality analysis of telephone exchange system that contains MDF, power supply, DP, FP, and DC.

With the help of Figure 7.2, we observe a graph of availability with respect to a function of time. The graph of reliability shows the sharply decrement from $t =$ 0 to 6, some very less increment from $t = 7$ to 11, and then constant. In Figure 7.3, we see a graph of reliability with respect to a function of time. With the help of graph of reliability we observe that the reliability of the system decreases with respect to increases in time. The Figure 7.4 shows the graph of MTTF as a function of various failure rates. With the help of a graph of MTTF, we observe that the MTTF with respect to failures of DP, DC, MDF, and power supply are near about same, but the MTTF with respect to failure of FP is the largest. The expected profit decreases due to the increments in service cost, the graph of Figure 7.5 explains this statement. This graph shows the variations by assuming the different values of service cost. Finally, Figures 7.6 and 7.7 show the graph of sensitivity analysis of reliability and MTTF. On observing the graph of Figure 7.6, we can say that the system is more sensitive with respect to failures of DC, power supply, and MDF, and the graph of Figure 7.7 shows the system is more sensitive with respect to failure of DP.

# References


[1] Flowers, T.H. (1952). Electronic telephone exchanges, Proceedings of the IEE-Part I: General, 99(119), 181–193.

[2] Palmer, R.W. (1955). Maintenance principles for automatic telephone exchange plant, Proceedings of the IEE-Part B: Radio and Electronic Engineering, 102(4), 453–468.

[3] Depp, W.A. and Townsend, M.A. (1964). An electronic-private-branch-exchange telephone switching system, IEEE Transactions on Communication and Electronics, 83(73), 329–331.

[4] Warman, J.B. and Bear, D. (1966). Trunking and traffic aspects of a sectionalised telephone exchange system. Electrical Engineers, Proceedings of the Institution of, 113(8), 1331–1343.

[5] Strandberg, K. and Ericsson, L.M. (1973). Reliability Prediction in Telephone Systems Engineering. 7 th InternationaL TeLetraffic Congress, StockhoLm.

[6] Malec, H.A. (1977). Reliability Optimization in Telephone Switching Systems Design, IEEE Transactions on Reliability, 26(3), 203–208.

[7] Baron, J., Traon, J.L., Riou, M. and Auregan, M. (1977). Contact Behavior of Telephone Relays and Connectors in Various Aggressive Environments, IEEE Transactions on Parts, Hybrids and Packaging, 13(1), 68–72.

[8] Tortorella, M. (1981). Cutoff calls and telephone equipment reliability, Bell Syst. Tech. J, 60(8), 1861–1890.

[9] Lelievre, A. and Goarin, R. (1987). The influence of climatic conditions on the reliability of electronic components in telephone exchanges, Quality and Reliability Engineering: International, 3(2), 93–98.

[10] Kolte, J. (1989). Cooling system BPA-105 for small telephone exchanges, Ericsson Review, 66(2), 58–63.

[11] Kanoun, K., de Bastos Martini, M.R. and de Souza, J.M. (1991). A method for software reliability analysis and prediction application to the TROPICO-R switching system, IEEE Transactions on Software Engineering, 17(4), 334–344.

[12] Fagerstrom, R. and Healy, J. (1993). The reliability of LEC telephone networks, IEEE Communications Magazine, 31(6), 44–48.

[13] Kuhn, D.R. (1997). Sources of failure in the public switched telephone network, Computers, 30(4), 31–36.

[14] Hellström, G., Enlund, S. and Paksoy, H. (2000). Direct Cooling of Telephone Switching Exchanges Using Borehole Heat Exchangers in Different Climates. In Proc. of 8th Int. Conf. on Thermal Energy Storage, Terrastock.

[15] Mirjalily, G., Bastani, P. and Almodarresi, S.M.T. (2008). Connecting telephone exchange centers over metro Ethernet networks by using Diffserv-MPLS. In Computer and Information Technology, 2008. ICCIT 2008. 11th International Conference on (pp. 708–713). IEEE.

[16] Mirjalily, G., Bastani, P. and Almodarresi, S.M.T. (2009). Performance evaluation of different QoS models for connecting telephone exchange centers over metro Ethernet networks, Journal of Networks, 4(10), 960–967.

[17] Manglik, M. and Ram, M. (2014). Stochastic modeling of a multi-state manufacturing system under three types of failures with perfect fault coverage, Journal of Engineering Science and Technology, 77–90.

[18] Nagiya, K. and Ram, M. (2014). Performance evaluation of a computer workstation under ring topology, Journal of Engineering Science and Technology, 91–103.

S.C. Malik and S.K. Chauhan

# 8 On use of Weibull failure laws for reliability measures of a series–parallel system

**Abstract:** Here, the reliability measures of a series-parallel system of (m, n) order have been obtained. The system has 'm' subsystems connected in series and each subsystem has 'n' components connected in parallel. The use of Weibull failure laws has been made to derive the expressions for mean time to system failure (MTSF) and reliability of the system. The expressions for these measures have also been obtained for the particular cases of the Weibull distribution i.e. for the Rayleigh and Exponential distributions. The results are derived for arbitrary values of the parameters related to number of components, operating time of the components and failure rate of the components. The behaviour of MTSF and reliability has been observed graphically for a particular system of (10, 10) order by considering all components identical in nature. The effect of operating time, scale and shape parameters on MTSF and reliability has also been shown graphically and numerically.

**Keywords:** structural design, reliability measures, arbitrary distribution, MTSF

## 8.1 Introduction

Structural designs of the components play an important role to develop reliable systems. The system designers and engineers have succeeded in identifying some better and effective structures of the components including series, parallel, series–parallel, and parallel–series. And, the parallel structure of the components has been suggested as the best one so far as reliability is concerned [1, 2]. But, the structures series–parallel and parallel–series [3, 4] involve a combination of several components and are very complex in nature. Therefore, it becomes necessary to know the effect of the number of components in the system, their qualities, and the structure in which they are arranged before developing the system. [5] developed reliability measures of a series–parallel system with some arbitrary distributions. [6] considered selective maintenance for multistate series–parallel system under different conditions. [7] discussed reliability equivalence factors of a series–parallel system in Weibull distribution. The purpose of the present chapter is to derive expressions for reliability and mean time to system failure (MTSF) of a series–parallel system of "*m*"

**S. C. Malik,** Department of Statistics, M. D. University, Rohtak, India
**S. K. Chauhan,** Department of Statistics, Shaheed Rajguru College of Applied Sciences for Women University of Delhi, New Delhi, India

https://doi.org/10.1515/9783110619058-008

subsystems each having "$n$" nonidentical components with Weibull failure laws. The expressions of these measures are also obtained for identical components. The values of MTSF and reliability of this system have been evaluated for arbitrary values of the number of components, operating time of the components, and failure rate of the components. The behavior of MTSF and reliability has been observed for monotonic failure nature of the components, that is,. by considering a particular case of Weibull failure laws. To make the study more effective, the trend of MTSF and Reliability has been observed numerically as well as graphically for a particular system of (10, 10) order under different operating time, scale, and shape parameters.

## 8.2 Notations

The following notations are used to represent different functions:

$R_S(t)$ = Reliability of the system, $R_i(t)$ = Reliability of the $i$th component
$R(t)$ = Reliability of identical components
$h(t)$ = Instantaneous failure rate of the system
$h_i(t)$ = Instantaneous failure rate of $i$th component, $\lambda$ = constant failure rate
$T$ = Lifetime of the system, $T_i$ = Lifetime of the $i$th component.
$m$ = Number of subsystems, $n$ = Number of components connected in series

## 8.3 System description

A series–parallel configuration of order $(m, n)$ is a system having "$m$" independent series of subsystems each has "$n$" components arranged in parallel network as shown in Figure 8.1:

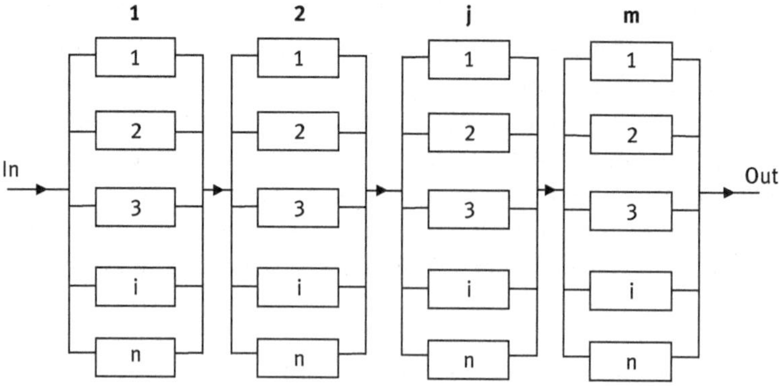


**Figure 8.1:** Series–parallel system of "$n$" components.

The system reliability at time "$t$" is given by

$$R_s(t) = \prod_{j=1}^{m}\left[1 - \prod_{i=1}^{n}(1 - R_i(t))\right] \tag{8.1}$$

where $1 - \prod_{i=1}^{n}(1 - R_i(t))$ is the reliability of a parallel system of "$n$" components.

Also, the MTSF is given by

$$\text{MTSF} = \int_0^{\infty} R(t)dt = \int_0^{\infty} \prod_{j=1}^{m}\left[1 - \prod_{i=1}^{n}(1 - R_i(t))\right] dt \tag{8.2}$$

When components in each subsystem are identical, that is, having the same reliability then, we have

$$R_s(t) = [1 - (1 - R(t))^n]^m$$

And,

$$\text{MTSF} = \int_0^{\infty} R(t)dt = \int_0^{\infty} [1 - (1 - R(t))^n]^m dt$$

## 8.4 Reliability measures

The reliability and MTSF of the series–parallel system have been obtained by considering Weibull distribution for failure rate of the components. The values of these reliability measures have also been evaluated for particular cases of the Weibull distribution.

## 8.5 Reliability measures by Weibull distribution

Suppose, failure rate of all the components are governed by Weibull failure laws, that is, $h_i(t) = \lambda_i t^{\beta_i}$

The components reliability is given by

$$R_i(t) = e^{-\int_0^t h_i(u)du} = e^{-\int_0^t \lambda_i u^{\beta_i} du} = e^{-\lambda_i \frac{t^{\beta_i+1}}{\beta_i+1}}$$

Therefore, the system reliability is given by

$$R_s(t) = \prod_{j=1}^{m}\left[1 - \prod_{i=1}^{n}\left(1 - e^{-\lambda_i \frac{t^{\beta_i+1}}{\beta_i+1}}\right)\right] \tag{8.3}$$

and

$$\text{MTSF} = \int_0^\infty R_s(t)dt = \int_0^\infty \prod_{j=1}^{m}\left[1-\prod_{i=1}^{n}\left(1-e^{-\frac{\lambda_i t^{\beta_i+1}}{\beta_i+1}}\right)\right]dt \tag{8.4}$$

For identical components, we have $h_i(t) = \lambda t^{\beta}$

The component reliability is given by $R_i(t) = e^{-\lambda\frac{t^{\beta+1}}{\beta+1}}$

Therefore, the system reliability is given by

$$R_s(t) = \left[1-\left(1-e^{-\lambda\frac{t^{\beta+1}}{\beta+1}}\right)^n\right]^m \tag{8.5}$$

and

$$\text{MTSF} = \int_0^\infty \left[1-\left(1-e^{-\lambda\frac{t^{\beta+1}}{\beta+1}}\right)^n\right]^m dt \tag{8.6}$$

## 8.6 Reliability measures for arbitrary values of the parameters

Reliability and MTSF of the system have been obtained for arbitrary values of the parameters associated with number of subsystems (m), number of components (n), scale parameter (λ), operating time of the component (t), and shape parameter (β). The results are shown numerically in the tables 8.1 to 8.5 and graphically in the figures 8.2 to 8.6.

**Table 8.1:** Reliability versus no. of subsystems (*m*) and components (*n*).

| $m$ | $n$ | $\lambda=0.01$, $\beta=0.1$, $t=10$ | $\lambda=0.02$, $\beta=0.1$, $t=10$ | $\lambda=0.03$, $\beta=0.1$, $t=10$ | $\lambda=0.04$, $\beta=0.1$, $t=10$ | $\lambda=0.05$, $\beta=0.1$, $t=10$ |
|---|---|---|---|---|---|---|
| 1 | 1 | 0.8918 | 0.7954 | 0.7094 | 0.6327 | 0.5643 |
| | 2 | 0.9883 | 0.9581 | 0.9155 | 0.8651 | 0.8101 |
| | 3 | 0.9987 | 0.9914 | 0.9754 | 0.9504 | 0.9173 |
| | 4 | 0.9998 | 0.9982 | 0.9929 | 0.9818 | 0.9639 |
| | 5 | 0.9999 | 0.9996 | 0.9979 | 0.9933 | 0.9843 |

**Table 8.1** (continued)

| $m$ | $n$ | $\lambda = 0.01$, $\beta = 0.1$, $t = 10$ | $\lambda = 0.02$, $\beta = 0.1$, $t = 10$ | $\lambda = 0.03$, $\beta = 0.1$, $t = 10$ | $\lambda = 0.04$, $\beta = 0.1$, $t = 10$ | $\lambda = 0.05$, $\beta = 0.1$, $t = 10$ |
|---|---|---|---|---|---|---|
| 2 | 1 | 0.7954 | 0.6327 | 0.5032 | 0.4003 | 0.3184 |
| | 2 | 0.9767 | 0.9180 | 0.8382 | 0.7483 | 0.6563 |
| | 3 | 0.9975 | 0.9829 | 0.9515 | 0.9033 | 0.8414 |
| | 4 | 0.9997 | 0.9965 | 0.9858 | 0.9639 | 0.9292 |
| | 5 | 0.99997 | 0.9993 | 0.9958 | 0.9867 | 0.9688 |
| 3 | 1 | 0.7093 | 0.5032 | 0.3570 | 0.2532 | 0.1796 |
| | 2 | 0.9653 | 0.8796 | 0.7674 | 0.6474 | 0.5317 |
| | 3 | 0.9962 | 0.9745 | 0.9282 | 0.8586 | 0.7718 |
| | 4 | 0.9996 | 0.9947 | 0.9787 | 0.9464 | 0.8957 |
| | 5 | 0.99996 | 0.9989 | 0.9938 | 0.9801 | 0.9536 |
| 4 | 1 | 0.6326 | 0.4003 | 0.2532 | 0.1602 | 0.1014 |
| | 2 | 0.9540 | 0.8428 | 0.7026 | 0.5600 | 0.4308 |
| | 3 | 0.9949 | 0.9662 | 0.9054 | 0.8160 | 0.7079 |
| | 4 | 0.9994 | 0.9930 | 0.9718 | 0.9291 | 0.8634 |
| | 5 | 0.99994 | 0.9986 | 0.9917 | 0.9735 | 0.9386 |
| 5 | 1 | 0.5643 | 0.3184 | 0.1796 | 0.1014 | 0.0572 |
| | 2 | 0.9429 | 0.8075 | 0.6433 | 0.4845 | 0.3490 |
| | 3 | 0.9937 | 0.9579 | 0.8832 | 0.7756 | 0.6493 |
| | 4 | 0.9993 | 0.9913 | 0.9648 | 0.9122 | 0.8323 |
| | 5 | 0.99993 | 0.9982 | 0.9897 | 0.9670 | 0.9239 |
| 6 | 1 | 0.5032 | 0.2532 | 0.1274 | 0.0641 | 0.0323 |
| | 2 | 0.9318 | 0.7737 | 0.5889 | 0.4191 | 0.2827 |
| | 3 | 0.9924 | 0.9497 | 0.8615 | 0.7371 | 0.5956 |
| | 4 | 0.9992 | 0.9895 | 0.9580 | 0.8956 | 0.8023 |
| | 5 | 0.99991 | 0.9978 | 0.9876 | 0.9605 | 0.9094 |

**Table 8.1** (continued)

| $m$ | $n$ | $\lambda = 0.01$, $\beta = 0.1$, $t = 10$ | $\lambda = 0.02$, $\beta = 0.1$, $t = 10$ | $\lambda = 0.03$, $\beta = 0.1$, $t = 10$ | $\lambda = 0.04$, $\beta = 0.1$, $t = 10$ | $\lambda = 0.05$, $\beta = 0.1$, $t = 10$ |
|---|---|---|---|---|---|---|
| 7 | 1 | 0.4488 | 0.2014 | 0.0904 | 0.0406 | 0.0182 |
| | 2 | 0.9209 | 0.7413 | 0.5392 | 0.3625 | 0.2290 |
| | 3 | 0.9912 | 0.9416 | 0.8403 | 0.7006 | 0.5463 |
| | 4 | 0.9990 | 0.9878 | 0.9511 | 0.8793 | 0.7733 |
| | 5 | 0.99989 | 0.9975 | 0.9856 | 0.9541 | 0.8951 |
| 8 | 1 | 0.4003 | 0.1602 | 0.0641 | 0.0257 | 0.0103 |
| | 2 | 0.9102 | 0.7103 | 0.4937 | 0.3136 | 0.1855 |
| | 3 | 0.9899 | 0.9335 | 0.8197 | 0.6659 | 0.5011 |
| | 4 | 0.9989 | 0.9861 | 0.9443 | 0.8633 | 0.7455 |
| | 5 | 0.9998 | 0.9971 | 0.9835 | 0.9477 | 0.8810 |
| 9 | 1 | 0.3570 | 0.1274 | 0.0455 | 0.0162 | 0.0058 |
| | 2 | 0.8995 | 0.6806 | 0.4520 | 0.2713 | 0.1503 |
| | 3 | 0.9887 | 0.9255 | 0.7996 | 0.6329 | 0.4597 |
| | 4 | 0.9988 | 0.9843 | 0.9376 | 0.8476 | 0.7186 |
| | 5 | 0.9998 | 0.9968 | 0.9815 | 0.9414 | 0.8672 |
| 10 | 1 | 0.3184 | 0.1014 | 0.0323 | 0.0103 | 0.0033 |
| | 2 | 0.8890 | 0.6521 | 0.4138 | 0.2347 | 0.1218 |
| | 3 | 0.9874 | 0.9176 | 0.7710 | 0.6015 | 0.4216 |
| | 4 | 0.9986 | 0.9826 | 0.9309 | 0.8322 | 0.6927 |
| | 5 | 0.9998 | 0.9964 | 0.9795 | 0.9351 | 0.8536 |

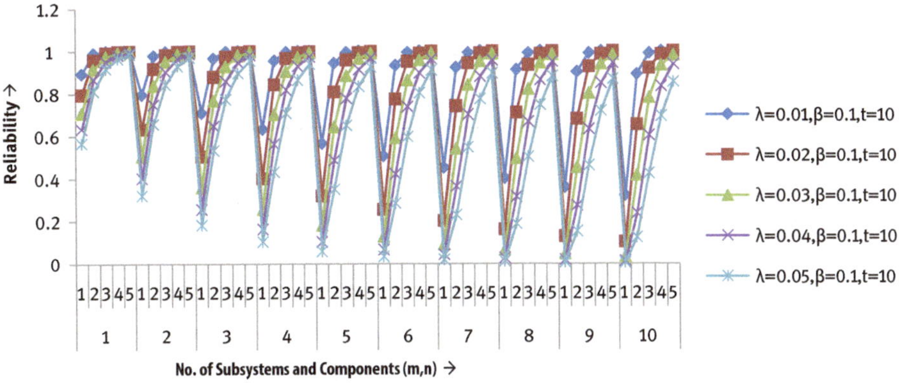


**Figure 8.2:** Reliability versus no. of subsystems (*m*) and components (*n*).

**Table 8.2:** MTSF versus no. of subsystems (*m*) and components (*n*).

| m | n | $\lambda = 0.01$, $\beta = 0.1$, $t = 10$ | $\lambda = 0.02$, $\beta = 0.1$, $t = 10$ | $\lambda = 0.03$, $\beta = 0.1$, $t = 10$ | $\lambda = 0.04$, $\beta = 0.1$, $t = 10$ | $\lambda = 0.05$, $\beta = 0.1$, $t = 10$ |
|---|---|---|---|---|---|---|
| 1 | 1 | 69.2308 | 36.8668 | 25.5007 | 19.6323 | 16.0277 |
| | 2 | 101.5948 | 54.1013 | 37.4218 | 28.8101 | 23.5204 |
| | 3 | 122.5927 | 65.2831 | 45.1563 | 34.7646 | 28.3816 |
| | 4 | 138.0929 | 73.5373 | 50.8657 | 39.1601 | 31.9701 |
| | 5 | 150.3592 | 80.0694 | 55.3839 | 42.6386 | 34.8099 |
| 2 | 1 | 36.8668 | 19.6323 | 13.5797 | 10.4546 | 8.5351 |
| | 2 | 65.0967 | 34.6653 | 23.9780 | 18.46 | 15.071 |
| | 3 | 84.6865 | 45.0973 | 31.1938 | 24.0152 | 19.6059 |
| | 4 | 99.5334 | 53.0036 | 36.6625 | 28.2255 | 23.0431 |
| | 5 | 111.4474 | 59.3480 | 41.0510 | 31.6040 | 25.8014 |
| 3 | 1 | 25.5007 | 13.5797 | 9.3930 | 7.2314 | 5.9037 |
| | 2 | 51.0045 | 27.1610 | 18.7872 | 14.4638 | 11.8081 |
| | 3 | 69.5870 | 37.0565 | 25.6319 | 19.7333 | 16.1102 |
| | 4 | 83.9388 | 44.6992 | 30.9184 | 23.8032 | 19.4328 |
| | 5 | 95.5704 | 50.8932 | 35.2028 | 27.1017 | 22.1256 |

**Table 8.2** (continued)

| m | n | $\lambda = 0.01$, $\beta = 0.1, t = 10$ | $\lambda = 0.02$, $\beta = 0.1, t = 10$ | $\lambda = 0.03$, $\beta = 0.1, t = 10$ | $\lambda = 0.04$, $\beta = 0.1, t = 10$ | $\lambda = 0.05$, $\beta = 0.1, t = 10$ |
|---|---|---|---|---|---|---|
| 4 | 1 | 19.6323 | 10.4546 | 7.2314 | 5.5673 | 4.5451 |
| | 2 | 43.1648 | 22.9862 | 15.8995 | 12.2406 | 9.9932 |
| | 3 | 60.9823 | 32.4743 | 22.4625 | 17.2933 | 14.1181 |
| | 4 | 74.9482 | 39.9114 | 27.6067 | 21.2537 | 17.3514 |
| | 5 | 86.3546 | 45.9856 | 31.8082 | 24.4883 | 19.9921 |
| 5 | 1 | 16.0277 | 8.5351 | 5.9037 | 4.54511 | 3.7106 |
| | 2 | 38.0426 | 20.2585 | 14.0128 | 10.7881 | 8.8073 |
| | 3 | 55.2491 | 29.4213 | 20.3507 | 15.6674 | 12.7908 |
| | 4 | 68.9012 | 36.6913 | 25.3793 | 19.5389 | 15.9514 |
| | 5 | 80.1222 | 42.6667 | 29.5125 | 22.7209 | 18.5492 |
| 6 | 1 | 13.5797 | 7.23145 | 5.0020 | 3.8509 | 3.1438 |
| | 2 | 34.3753 | 18.3055 | 12.6619 | 9.7481 | 7.9583 |
| | 3 | 51.0757 | 27.1989 | 18.8134 | 14.4840 | 11.8246 |
| | 4 | 64.4644 | 34.3286 | 23.7451 | 18.2807 | 14.9243 |
| | 5 | 75.5285 | 40.2205 | 27.8205 | 21.4182 | 17.4857 |
| 7 | 1 | 11.8040 | 6.28586 | 4.3479 | 3.3473 | 2.7327 |
| | 2 | 31.5894 | 16.8220 | 11.6358 | 8.9581 | 7.3133 |
| | 3 | 47.8596 | 25.4862 | 17.6288 | 13.5719 | 11.0801 |
| | 4 | 61.0220 | 32.4954 | 22.4771 | 17.3045 | 14.1273 |
| | 5 | 71.9501 | 38.3149 | 26.5024 | 20.4035 | 16.6573 |
| 8 | 1 | 10.4546 | 5.56730 | 3.8509 | 2.96470 | 2.4204 |
| | 2 | 29.3835 | 15.6473 | 10.8232 | 8.3325 | 6.8026 |
| | 3 | 45.2806 | 24.1128 | 16.6788 | 12.8406 | 10.4830 |
| | 4 | 58.2447 | 31.0165 | 21.4541 | 16.5169 | 13.4843 |
| | 5 | 69.0532 | 36.7722 | 25.4353 | 19.5820 | 15.9866 |

**Table 8.2** (continued)

| m | n | $\lambda=0.01$, $\beta=0.1, t=10$ | $\lambda=0.02$, $\beta=0.1, t=10$ | $\lambda=0.03$, $\beta=0.1, t=10$ | $\lambda=0.04$, $\beta=0.1, t=10$ | $\lambda=0.05$, $\beta=0.1, t=10$ |
|---|---|---|---|---|---|---|
| 9 | 1 | 9.39304 | 5.00199 | 3.4599 | 2.6637 | 2.1746 |
| | 2 | 27.5825 | 14.6882 | 10.1598 | 7.8218 | 6.3857 |
| | 3 | 43.1507 | 22.9787 | 15.8943 | 12.2366 | 9.9899 |
| | 4 | 55.9389 | 29.7886 | 20.6048 | 15.8631 | 12.9505 |
| | 5 | 66.6405 | 35.4875 | 24.5466 | 18.8978 | 15.4281 |
| 10 | 1 | 8.53510 | 4.5451 | 3.1438 | 2.4204 | 1.9760 |
| | 2 | 26.0769 | 13.8865 | 9.6052 | 7.3948 | 6.0371 |
| | 3 | 41.3518 | 22.0207 | 15.2317 | 11.7265 | 9.5734 |
| | 4 | 53.9819 | 28.7465 | 19.8839 | 15.3081 | 12.4974 |
| | 5 | 64.5871 | 34.3939 | 23.7902 | 18.3155 | 14.9527 |

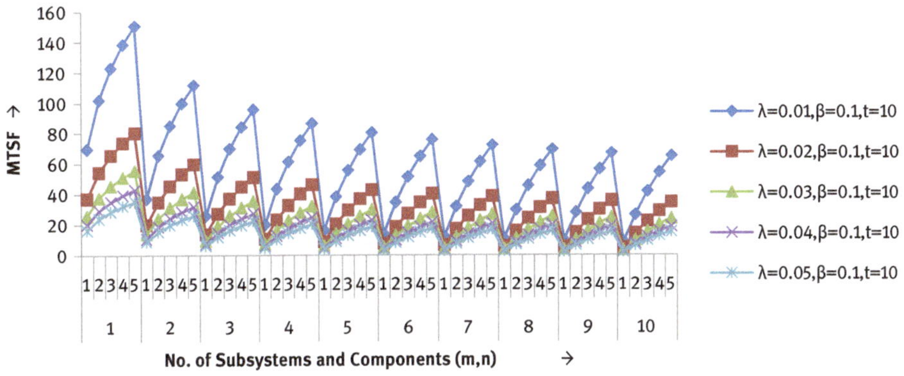


**Figure 8.3:** MTSF versus no. of subsystems (*m*) and components (*n*).

**Table 8.3:** Reliability versus no. of subsystems (*m*) and components (*n*)

| m | n | $\beta=0.1$, $\lambda=0.01, t=10$ | $\beta=0.2$, $\lambda=0.01, t=10$ | $\beta=0.3$, $\lambda=0.01, t=10$ | $\beta=0.4$, $\lambda=0.01, t=10$ | $\beta=0.5$, $\lambda=0.01, t=10$ |
|---|---|---|---|---|---|---|
| 1 | 1 | 0.8918 | 0.8763 | 0.8577 | 0.8357 | 0.8099 |
| | 2 | 0.9883 | 0.9847 | 0.9797 | 0.9730 | 0.9639 |
| | 3 | 0.9987 | 0.9981 | 0.9971 | 0.9956 | 0.9931 |
| | 4 | 0.9999 | 0.9998 | 0.9996 | 0.9993 | 0.9987 |
| | 5 | 0.99998 | 0.99997 | 0.99994 | 0.9998 | 0.9997 |
| 2 | 1 | 0.7954 | 0.7678 | 0.7357 | 0.6985 | 0.6560 |
| | 2 | 0.9767 | 0.9696 | 0.9599 | 0.9468 | 0.9290 |
| | 3 | 0.9975 | 0.9962 | 0.9942 | 0.9911 | 0.9863 |
| | 4 | 0.9997 | 0.9995 | 0.9992 | 0.9985 | 0.9974 |
| | 5 | 0.99997 | 0.99994 | 0.99988 | 0.9998 | 0.9995 |
| 3 | 1 | 0.7094 | 0.6728 | 0.6310 | 0.5838 | 0.5313 |
| | 2 | 0.9653 | 0.9548 | 0.9405 | 0.9212 | 0.8955 |
| | 3 | 0.9962 | 0.9943 | 0.9914 | 0.9868 | 0.9795 |
| | 4 | 0.9996 | 0.9993 | 0.9988 | 0.9978 | 0.9961 |
| | 5 | 0.99996 | 0.99991 | 0.9998 | 0.9996 | 0.9992 |
| 4 | 1 | 0.6327 | 0.5896 | 0.5412 | 0.4879 | 0.4303 |
| | 2 | 0.9540 | 0.9402 | 0.9214 | 0.8964 | 0.8631 |
| | 3 | 0.9949 | 0.9924 | 0.9885 | 0.9824 | 0.9728 |
| | 4 | 0.9994 | 0.9991 | 0.9984 | 0.9971 | 0.9948 |
| | 5 | 0.99994 | 0.9998 | 0.9998 | 0.9995 | 0.9990 |
| 5 | 1 | 0.5643 | 0.5167 | 0.4642 | 0.4077 | 0.3485 |
| | 2 | 0.9429 | 0.9258 | 0.9028 | 0.8722 | 0.8319 |
| | 3 | 0.9937 | 0.9906 | 0.9857 | 0.9780 | 0.9661 |
| | 4 | 0.9993 | 0.9988 | 0.9979 | 0.9964 | 0.9935 |
| | 5 | 0.99992 | 0.9998 | 0.9997 | 0.9994 | 0.9988 |

**Table 8.3** (continued)

| m | n | $\beta = 0.1$, $\lambda = 0.01$, $t = 10$ | $\beta = 0.2$, $\lambda = 0.01$, $t = 10$ | $\beta = 0.3$, $\lambda = 0.01$, $t = 10$ | $\beta = 0.4$, $\lambda = 0.01$, $t = 10$ | $\beta = 0.5$, $\lambda = 0.01$, $t = 10$ |
|---|---|---|---|---|---|---|
| 6 | 1 | 0.5032 | 0.4527 | 0.3982 | 0.3408 | 0.2823 |
| | 2 | 0.9318 | 0.9116 | 0.8845 | 0.8487 | 0.8019 |
| | 3 | 0.9924 | 0.9887 | 0.9828 | 0.9737 | 0.9595 |
| | 4 | 0.9992 | 0.9986 | 0.9975 | 0.9956 | 0.9922 |
| | 5 | 0.99991 | 0.9998 | 0.9996 | 0.9993 | 0.9985 |
| 7 | 1 | 0.4488 | 0.3967 | 0.3415 | 0.2848 | 0.2286 |
| | 2 | 0.9209 | 0.8976 | 0.8666 | 0.8258 | 0.7729 |
| | 3 | 0.9912 | 0.9868 | 0.9800 | 0.9694 | 0.9529 |
| | 4 | 0.9990 | 0.9984 | 0.9971 | 0.9949 | 0.9909 |
| | 5 | 0.9998 | 0.9998 | 0.9996 | 0.9992 | 0.9982 |
| 8 | 1 | 0.4002 | 0.3476 | 0.2929 | 0.2380 | 0.1851 |
| | 2 | 0.9102 | 0.8839 | 0.8491 | 0.8035 | 0.7450 |
| | 3 | 0.9899 | 0.9849 | 0.9772 | 0.9651 | 0.9464 |
| | 4 | 0.9989 | 0.9981 | 0.9967 | 0.9942 | 0.9896 |
| | 5 | 0.99988 | 0.9998 | 0.9995 | 0.9990 | 0.9980 |
| 9 | 1 | 0.3570 | 0.3046 | 0.2512 | 0.1989 | 0.1410 |
| | 2 | 0.8995 | 0.8704 | 0.8319 | 0.7818 | 0.7181 |
| | 3 | 0.9887 | 0.9831 | 0.9744 | 0.9608 | 0.9399 |
| | 4 | 0.9988 | 0.9979 | 0.9963 | 0.9935 | 0.9883 |
| | 5 | 0.99986 | 0.9997 | 0.9995 | 0.9989 | 0.9978 |
| 10 | 1 | 0.3184 | 0.2669 | 0.2155 | 0.1663 | 0.1214 |
| | 2 | 0.8890 | 0.8570 | 0.8150 | 0.7607 | 0.6921 |
| | 3 | 0.9874 | 0.9812 | 0.9716 | 0.9566 | 0.9334 |
| | 4 | 0.9986 | 0.9976 | 0.9959 | 0.9927 | 0.9870 |
| | 5 | 0.9998 | 0.9997 | 0.9994 | 0.9988 | 0.9975 |

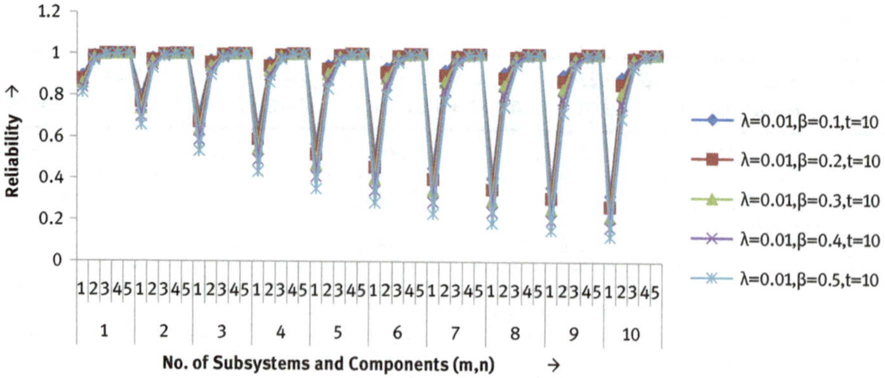


**Figure 8.4:** Reliability versus no. of subsystems (*m*) and components (*n*).

**Table 8.4:** MTSF versus no. of subsystems (*m*) and components (*n*).

| m | n | $\beta=0.1$, $\lambda=0.01$, $t=10$ | $\beta=0.2$, $\lambda=0.01$, $t=10$ | $\beta=0.3$, $\lambda=0.01$, $t=10$ | $\beta=0.4$, $\lambda=0.01$, $t=10$ | $\beta=0.4$, $\lambda=0.01$, $t=10$ |
|---|---|---|---|---|---|---|
| 1 | 1 | 69.2308 | 50.8255 | 39.0466 | 31.0934 | 25.4855 |
| | 2 | 101.5948 | 73.1262 | 55.1834 | 43.2352 | 34.9161 |
| | 3 | 122.5927 | 87.2481 | 65.1816 | 50.6116 | 40.5440 |
| | 4 | 138.0929 | 97.5282 | 72.3708 | 55.8576 | 44.5075 |
| | 5 | 150.3592 | 105.5872 | 77.9602 | 59.9066 | 47.5466 |
| 2 | 1 | 36.8668 | 28.5249 | 22.9098 | 18.9517 | 16.0548 |
| | 2 | 65.0967 | 48.7241 | 37.9959 | 30.6128 | 25.3247 |
| | 3 | 84.6865 | 62.2936 | 47.8428 | 38.0308 | 31.0868 |
| | 4 | 99.5334 | 72.3941 | 55.0578 | 43.3911 | 35.1993 |
| | 5 | 111.4474 | 80.4033 | 60.7199 | 47.5595 | 38.3719 |
| 3 | 1 | 25.5007 | 20.3461 | 16.7713 | 14.1863 | 12.2521 |
| | 2 | 51.0045 | 38.9964 | 30.9582 | 25.3251 | 21.2272 |
| | 3 | 69.5870 | 52.0807 | 40.5842 | 32.6618 | 26.9835 |
| | 4 | 83.9388 | 61.9800 | 47.7375 | 38.0285 | 31.1358 |
| | 5 | 95.5704 | 69.8955 | 53.3909 | 42.2269 | 34.3551 |

**Table 8.4** (continued)

| m | n | $\beta=0.1$, $\lambda=0.01, t=10$ | $\beta=0.2$, $\lambda=0.01, t=10$ | $\beta=0.3$, $\lambda=0.01, t=10$ | $\beta=0.4$, $\lambda=0.01, t=10$ | $\beta=0.4$, $\lambda=0.01, t=10$ |
|---|---|---|---|---|---|---|
| 4 | 1 | 19.6323 | 16.0090 | 13.4419 | 11.5512 | 10.1139 |
| | 2 | 43.1648 | 33.4832 | 26.9067 | 22.2403 | 18.8091 |
| | 3 | 60.9823 | 46.1712 | 36.3299 | 29.4805 | 24.5292 |
| | 4 | 74.9482 | 55.8949 | 43.4117 | 34.8292 | 28.6914 |
| | 5 | 86.3546 | 63.7214 | 49.0403 | 39.0337 | 31.9316 |
| 5 | 1 | 16.0277 | 13.2925 | 11.3218 | 9.8493 | 8.7159 |
| | 2 | 38.0426 | 29.8335 | 24.1949 | 20.1560 | 17.1620 |
| | 3 | 55.2491 | 42.1918 | 33.4395 | 27.3026 | 22.8381 |
| | 4 | 68.9012 | 51.7643 | 40.4526 | 32.6261 | 26.9986 |
| | 5 | 80.1222 | 59.5112 | 46.0528 | 36.8279 | 30.249 |
| 6 | 1 | 13.5797 | 11.4188 | 9.84023 | 8.6466 | 7.7184 |
| | 2 | 34.3753 | 27.1938 | 22.2168 | 18.6246 | 15.9441 |
| | 3 | 51.0757 | 39.2716 | 31.3040 | 25.6843 | 21.5751 |
| | 4 | 64.4644 | 48.7125 | 38.2535 | 30.9807 | 25.7288 |
| | 5 | 75.5285 | 56.3885 | 43.8253 | 35.1759 | 28.9839 |
| 7 | 1 | 11.8040 | 10.0423 | 8.7399 | 7.7451 | 6.9646 |
| | 2 | 31.5894 | 25.1721 | 20.6912 | 17.4365 | 14.9944 |
| | 3 | 47.8596 | 37.0065 | 29.6386 | 24.4162 | 20.5815 |
| | 4 | 61.0220 | 46.3314 | 36.5296 | 29.6857 | 24.7261 |
| | 5 | 71.9501 | 53.9439 | 42.0742 | 33.8725 | 27.9828 |
| 8 | 1 | 10.4546 | 8.98477 | 7.8868 | 7.0405 | 6.3714 |
| | 2 | 29.3835 | 23.5601 | 19.4678 | 16.4789 | 14.2257 |
| | 3 | 45.2806 | 35.1803 | 28.2896 | 23.3851 | 19.7708 |
| | 4 | 58.2447 | 44.4015 | 35.1269 | 28.6285 | 23.9051 |
| | 5 | 69.0532 | 51.9567 | 40.6457 | 32.8061 | 27.1616 |

**Table 8.4** (continued)

| m | n | $\beta=0.1$, $\lambda=0.01$, $t=10$ | $\beta=0.2$, $\lambda=0.01$, $t=10$ | $\beta=0.3$, $\lambda=0.01$, $t=10$ | $\beta=0.4$, $\lambda=0.01$, $t=10$ | $\beta=0.4$, $\lambda=0.01$, $t=10$ |
|---|---|---|---|---|---|---|
| 9 | 1 | 9.39304 | 8.14479 | 7.2036 | 6.4724 | 5.8902 |
| | 2 | 27.5825 | 22.2361 | 18.4578 | 15.6850 | 13.5861 |
| | 3 | 43.1507 | 33.6651 | 27.1660 | 22.5234 | 19.0914 |
| | 4 | 55.9389 | 42.7928 | 33.9538 | 27.7419 | 23.2148 |
| | 5 | 66.6405 | 50.2958 | 39.4484 | 31.9100 | 26.4700 |
| 10 | 1 | 8.5351 | 7.4602 | 6.6428 | 6.0032 | 5.4907 |
| | 2 | 26.0769 | 21.1235 | 17.6053 | 15.0124 | 13.0424 |
| | 3 | 41.3518 | 32.3801 | 26.2099 | 21.7880 | 18.5101 |
| | 4 | 53.9819 | 41.4228 | 32.9519 | 26.9827 | 22.6225 |
| | 5 | 64.5871 | 48.8779 | 38.4235 | 31.1414 | 25.8756 |

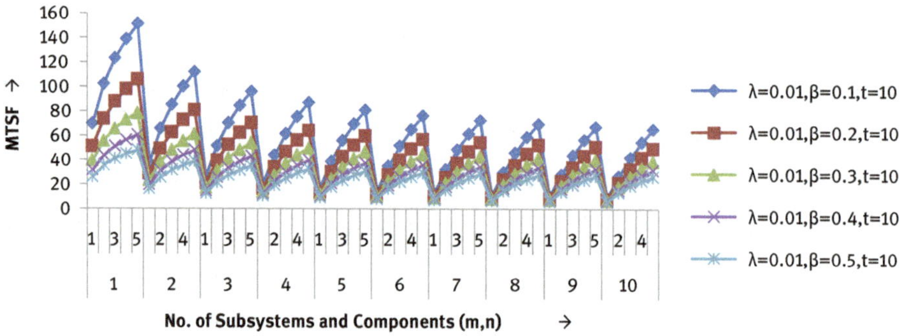


**Figure 8.5:** MTSF versus no. of subsystems (*m*) and components (*n*).

**Table 8.5:** Reliability versus no. of subsystems ($m$) and components (n).

| m | n | $t=5$, $\lambda=0.01, \beta=0.1$ | $t=10$, $\lambda=0.01, \beta=0.1$ | $t=15$, $\lambda=0.01, \beta=0.1$ | $t=20$, $\lambda=0.01, \beta=0.1$ | $t=25$, $\lambda=0.01, \beta=0.1$ |
|---|---|---|---|---|---|---|
| 1 | 1 | 0.9480 | 0.8918 | 0.8363 | 0.7824 | 0.7308 |
| | 2 | 0.9973 | 0.9883 | 0.9732 | 0.9527 | 0.9275 |
| | 3 | 0.9998 | 0.9987 | 0.9956 | 0.9897 | 0.9805 |
| | 4 | 0.999993 | 0.9998 | 0.9993 | 0.9978 | 0.9947 |
| | 5 | 0.9999996 | 0.99998 | 0.9998 | 0.9995 | 0.9986 |
| 2 | 1 | 0.8987 | 0.7954 | 0.6994 | 0.6122 | 0.5341 |
| | 2 | 0.9946 | 0.9767 | 0.9471 | 0.9076 | 0.8603 |
| | 3 | 0.9997 | 0.9975 | 0.9912 | 0.9795 | 0.9614 |
| | 4 | 0.99998 | 0.9997 | 0.9986 | 0.9955 | 0.9895 |
| | 5 | 0.999999 | 0.99997 | 0.9998 | 0.9990 | 0.9972 |
| 3 | 1 | 0.8520 | 0.7094 | 0.5849 | 0.4790 | 0.3903 |
| | 2 | 0.9919 | 0.9653 | 0.9217 | 0.8646 | 0.7980 |
| | 3 | 0.9996 | 0.9962 | 0.9869 | 0.9694 | 0.9426 |
| | 4 | 0.99998 | 0.9996 | 0.9978 | 0.9933 | 0.9843 |
| | 5 | 0.999999 | 0.99996 | 0.9996 | 0.9985 | 0.9958 |
| 4 | 1 | 0.8077 | 0.6327 | 0.4891 | 0.3748 | 0.2853 |
| | 2 | 0.9892 | 0.9540 | 0.8970 | 0.8237 | 0.7402 |
| | 3 | 0.9994 | 0.9949 | 0.9826 | 0.9594 | 0.9242 |
| | 4 | 0.99997 | 0.9994 | 0.9971 | 0.9911 | 0.9792 |
| | 5 | 0.999998 | 0.99994 | 0.9995 | 0.9980 | 0.9944 |
| 5 | 1 | 0.7657 | 0.5643 | 0.4091 | 0.2933 | 0.2085 |
| | 2 | 0.9865 | 0.9429 | 0.8730 | 0.7847 | 0.6866 |
| | 3 | 0.9993 | 0.9937 | 0.9782 | 0.9496 | 0.9062 |
| | 4 | 0.99996 | 0.9993 | 0.9964 | 0.9888 | 0.9740 |
| | 5 | 0.999998 | 0.99993 | 0.9994 | 0.9976 | 0.9929 |

**Table 8.5** (continued)

| m | n | $t=5$, $\lambda=0.01, \beta=0.1$ | $t=10$, $\lambda=0.01, \beta=0.1$ | $t=15$, $\lambda=0.01, \beta=0.1$ | $t=20$, $\lambda=0.01, \beta=0.1$ | $t=25$, $\lambda=0.01, \beta=0.1$ |
|---|---|---|---|---|---|---|
| 6 | 1 | 0.7259 | 0.5032 | 0.3421 | 0.2295 | 0.1524 |
| | 2 | 0.9839 | 0.9318 | 0.8496 | 0.7476 | 0.6368 |
| | 3 | 0.9991 | 0.9924 | 0.9740 | 0.9398 | 0.8885 |
| | 4 | 0.99996 | 0.9992 | 0.9957 | 0.9866 | 0.9689 |
| | 5 | 0.999998 | 0.99991 | 0.9993 | 0.9971 | 0.9915 |
| 7 | 1 | 0.6881 | 0.4488 | 0.2861 | 0.1795 | 0.1113 |
| | 2 | 0.9812 | 0.9209 | 0.8268 | 0.7122 | 0.5907 |
| | 3 | 0.9990 | 0.9912 | 0.9697 | 0.9301 | 0.8712 |
| | 4 | 0.99995 | 0.9990 | 0.9950 | 0.9844 | 0.9638 |
| | 5 | 0.999997 | 0.99989 | 0.9992 | 0.9966 | 0.9901 |
| 8 | 1 | 0.6524 | 0.4003 | 0.2393 | 0.1405 | 0.0814 |
| | 2 | 0.9786 | 0.9102 | 0.8047 | 0.6785 | 0.5479 |
| | 3 | 0.9989 | 0.9899 | 0.9654 | 0.9205 | 0.8542 |
| | 4 | 0.99994 | 0.9989 | 0.9943 | 0.9822 | 0.9588 |
| | 5 | 0.999997 | 0.9998 | 0.9991 | 0.9961 | 0.9887 |
| 9 | 1 | 0.6184 | 0.3570 | 0.2001 | 0.1099 | 0.0595 |
| | 2 | 0.9759 | 0.8995 | 0.7831 | 0.6464 | 0.5082 |
| | 3 | 0.9987 | 0.9887 | 0.9612 | 0.9111 | 0.8376 |
| | 4 | 0.99993 | 0.9988 | 0.9935 | 0.9800 | 0.9537 |
| | 5 | 0.999997 | 0.9998 | 0.9989 | 0.9956 | 0.9873 |
| 10 | 1 | 0.5863 | 0.3184 | 0.1673 | 0.0860 | 0.0435 |
| | 2 | 0.9733 | 0.8890 | 0.7621 | 0.6158 | 0.4714 |
| | 3 | 0.9986 | 0.9874 | 0.9570 | 0.9017 | 0.8212 |
| | 4 | 0.99993 | 0.9986 | 0.9928 | 0.9778 | 0.9487 |
| | 5 | 0.999996 | 0.9998 | 0.9988 | 0.9951 | 0.9860 |

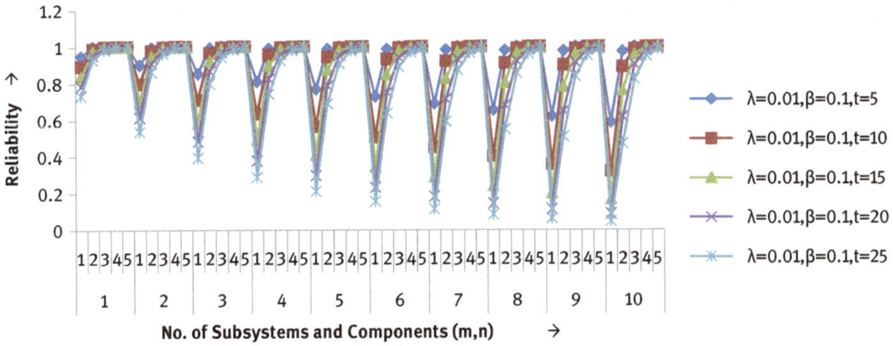


**Figure 8.6:** Reliability versus no. of subsystems (*m*) and components (*n*).

## 8.7 Particular cases of Weibull distribution Rayleigh distribution

The Rayleigh distribution is a special case of Weibull distribution with the shape parameter $\beta = 1$. The component reliability in this case is given by

$$R_i(t) = e^{-\int_0^t h_i(u)du} = e^{-\int_0^t \lambda_i u du} = e^{\frac{-\lambda_i t^2}{2}} \tag{8.7}$$

where $h_i(t) = \lambda_i t$. Therefore, the system reliability is given by

$$R_s(t) = \prod_{j=1}^{m} \left\{ 1 - \prod_{i=1}^{n} \left( 1 - e^{-\lambda_i \frac{t^2}{2}} \right) \right\} \tag{8.8}$$

and

$$\text{MTSF} = \int_0^\infty R_s(t)dt = \int_0^\infty \prod_{j=1}^{m} \left\{ 1 - \prod_{i=1}^{n} \left( 1 - e^{-\lambda_i \frac{t^2}{2}} \right) \right\} dt \tag{8.9}$$

For identical components, we have $h_i(t) = \lambda t$

The component reliability is given by $R_i(t) = e^{-\lambda \frac{t^2}{2}}$

Then the system reliability is given by

$$R_s(t) = \left[ 1 - \left( 1 - e^{-\lambda \frac{t^2}{2}} \right)^n \right]^m \text{ and MTSF} = \int_0^\infty \left[ 1 - \left( 1 - e^{-\lambda \frac{t^2}{2}} \right)^n \right]^m dt$$

## 8.8 Reliability measures for arbitrary values of the parameters

Reliability and MTSF of the system have been obtained for arbitrary values of the parameters associated with number of subsystems (m), number of components (n), failure rate (λ) and operating time of the components (t). The results are shown numerically in the tables 8.6 to 8.8 and graphically in figures 8.7 to 8.9.

**Table 8.6:** Reliability versus no. of subsystems (*m*) and components (*n*).

| m | n | $\lambda = 0.01, t = 10$ | $\lambda = 0.02, t = 10$ | $\lambda = 0.03, t = 10$ | $\lambda = 0.04, t = 10$ | $\lambda = 0.05, t = 10$ |
|---|---|---|---|---|---|---|
| 1 | 1 | 0.6065 | 0.3679 | 0.2231 | 0.1353 | 0.0821 |
| | 2 | 0.8452 | 0.6004 | 0.3965 | 0.2523 | 0.1574 |
| | 3 | 0.9391 | 0.7474 | 0.5311 | 0.3535 | 0.2266 |
| | 4 | 0.9760 | 0.8403 | 0.63575 | 0.4410 | 0.2901 |
| | 5 | 0.9906 | 0.8991 | 0.7170 | 0.5167 | 0.3483 |
| 2 | 1 | 0.3679 | 0.1353 | 0.0499 | 0.0183 | 0.0067 |
| | 2 | 0.7143 | 0.3605 | 0.1572 | 0.0637 | 0.0248 |
| | 3 | 0.8819 | 0.5586 | 0.2821 | 0.1250 | 0.0513 |
| | 4 | 0.9526 | 0.7062 | 0.4042 | 0.1945 | 0.0841 |
| | 5 | 0.9812 | 0.8083 | 0.5141 | 0.2669 | 0.1213 |
| 3 | 1 | 0.2231 | 0.0498 | 0.0111 | 0.0025 | 0.00055 |
| | 2 | 0.6037 | 0.2164 | 0.0623 | 0.0161 | 0.0039 |
| | 3 | 0.8281 | 0.4175 | 0.1498 | 0.0442 | 0.0116 |
| | 4 | 0.9298 | 0.5934 | 0.2570 | 0.0858 | 0.0244 |
| | 5 | 0.9720 | 0.7267 | 0.3686 | 0.1379 | 0.0423 |
| 4 | 1 | 0.1353 | 0.0183 | 0.0025 | 0.00033 | 0.00004 |
| | 2 | 0.5103 | 0.1299 | 0.0247 | 0.0041 | 0.0006 |
| | 3 | 0.7771 | 0.3121 | 0.0796 | 0.0156 | 0.0026 |
| | 4 | 0.9075 | 0.4987 | 0.1634 | 0.0378 | 0.0071 |
| | 5 | 0.9628 | 0.6534 | 0.2643 | 0.0713 | 0.0147 |

**Table 8.6** (continued)

| m | n | $\lambda = 0.01, t = 10$ | $\lambda = 0.02, t = 10$ | $\lambda = 0.03, t = 10$ | $\lambda = 0.04, t = 10$ | $\lambda = 0.05, t = 10$ |
|---|---|---|---|---|---|---|
| 5 | 1 | 0.0821 | 0.0067 | 0.00055 | 0.000045 | 0.0000037 |
| | 2 | 0.4313 | 0.0780 | 0.0098 | 0.0010 | 0.000097 |
| | 3 | 0.7303 | 0.2332 | 0.0423 | 0.0055 | 0.0006 |
| | 4 | 0.8858 | 0.4190 | 0.1039 | 0.0167 | 0.0020 |
| | 5 | 0.9537 | 0.5875 | 0.1895 | 0.0368 | 0.0051 |
| 6 | 1 | 0.0498 | 0.0025 | 0.00012 | 0.0000061 | 0.00000031 |
| | 2 | 0.3645 | 0.0468 | 0.0039 | 0.0003 | 0.000015 |
| | 3 | 0.6858 | 0.1743 | 0.0224 | 0.0020 | 0.00013 |
| | 4 | 0.8645 | 0.3521 | 0.0660 | 0.0073 | 0.00059 |
| | 5 | 0.9447 | 0.5282 | 0.1359 | 0.0190 | 0.00179 |
| 7 | 1 | 0.0302 | 0.0009 | 0.000027 | 0.0000008 | 0.000000025 |
| | 2 | 0.3081 | 0.0281 | 0.0015 | 0.000065 | 0.0000024 |
| | 3 | 0.6441 | 0.1303 | 0.0119 | 0.00069 | 0.000031 |
| | 4 | 0.8438 | 0.2959 | 0.0420 | 0.0032 | 0.00017 |
| | 5 | 0.9358 | 0.4749 | 0.0974 | 0.0098 | 0.00062 |
| 8 | 1 | 0.0183 | 0.0003 | 0.000006 | 0.00000011 | 0.0000000021 |
| | 2 | 0.2604 | 0.0169 | 0.00061 | 0.000016 | 0.00000037 |
| | 3 | 0.6048 | 0.0974 | 0.0063 | 0.0002 | 0.0000069 |
| | 4 | 0.8236 | 0.2487 | 0.0267 | 0.0014 | 0.00005 |
| | 5 | 0.9270 | 0.4269 | 0.0699 | 0.0051 | 0.00022 |
| 9 | 1 | 0.0111 | 0.00012 | 0.0000014 | 0.000000015 | 0.00000000017 |
| | 2 | 0.2201 | 0.0101 | 0.00024 | 0.0000041 | 0.000000059 |
| | 3 | 0.5680 | 0.0728 | 0.0034 | 0.000086 | 0.0000016 |
| | 4 | 0.8038 | 0.2090 | 0.0170 | 0.00063 | 0.000014 |
| | 5 | 0.9182 | 0.3838 | 0.0501 | 0.0026 | 0.000075 |

**Table 8.6** (continued)

| m | n | $\lambda=0.01, t=10$ | $\lambda=0.02, t=10$ | $\lambda=0.03, t=10$ | $\lambda=0.04, t=10$ | $\lambda=0.05, t=10$ |
|---|---|---|---|---|---|---|
| 10 | 1 | 0.0067 | 0.00004 | 0.0000003 | 0.0000000021 | 0.000000000014 |
| | 2 | 0.1860 | 0.0061 | 0.000096 | 0.0000011 | 0.0000000093 |
| | 3 | 0.5334 | 0.0544 | 0.0018 | 0.00003 | 0.00000036 |
| | 4 | 0.7846 | 0.1756 | 0.0108 | 0.0003 | 0.000004 |
| | 5 | 0.9096 | 0.3451 | 0.0359 | 0.0013 | 0.00003 |

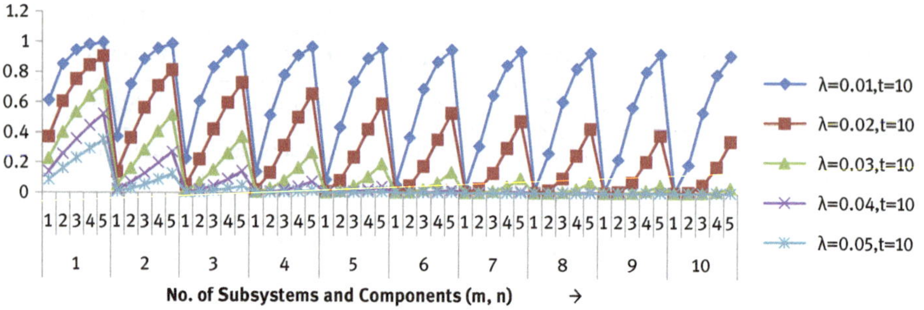


**Figure 8.7:** Reliability versus no. of subsystems (*m*) and components (*n*).

**Table 8.7:** MTSF vs no. of subsystems (*m*) and components (*n*).

| m | n | $\lambda=0.01, t=10$ | $\lambda=0.02, t=10$ | $\lambda=0.03, t=10$ | $\lambda=0.04, t=10$ | $\lambda=0.05, t=10$ |
|---|---|---|---|---|---|---|
| 1 | 1 | 12.5331 | 8.8623 | 7.2360 | 6.2666 | 5.6050 |
| | 2 | 16.2040 | 11.4580 | 9.3554 | 8.1021 | 7.2466 |
| | 3 | 18.2486 | 12.9037 | 10.536 | 9.1243 | 8.1610 |
| | 4 | 19.6364 | 13.8850 | 11.3371 | 9.8182 | 8.7817 |
| | 5 | 20.6753 | 14.6196 | 11.9369 | 10.3376 | 9.2463 |

**Table 8.7** (continued)

| m | n | $\lambda = 0.01, t = 10$ | $\lambda = 0.02, t = 10$ | $\lambda = 0.03, t = 10$ | $\lambda = 0.04, t = 10$ | $\lambda = 0.05, t = 10$ |
|---|---|---|---|---|---|---|
| 2 | 1 | 8.86227 | 6.2666 | 5.1166 | 4.4311 | 3.9633 |
| | 2 | 12.7716 | 9.0309 | 7.3737 | 6.3858 | 5.7116 |
| | 3 | 14.9974 | 10.6048 | 8.6588 | 7.4987 | 6.7071 |
| | 4 | 16.5152 | 11.6780 | 9.5351 | 8.2576 | 7.3858 |
| | 5 | 17.6522 | 12.4820 | 10.1915 | 8.8261 | 7.8943 |
| 3 | 1 | 7.23601 | 5.1166 | 4.1777 | 3.6180 | 3.2360 |
| | 2 | 11.2026 | 7.9214 | 6.4678 | 5.6013 | 5.0099 |
| | 3 | 13.5043 | 9.5490 | 7.7967 | 6.7521 | 6.0393 |
| | 4 | 15.0817 | 10.6644 | 8.7074 | 7.5408 | 6.7447 |
| | 5 | 16.2654 | 11.5014 | 9.3908 | 8.1327 | 7.2741 |
| 4 | 1 | 6.26657 | 4.4311 | 3.6180 | 3.1333 | 2.8025 |
| | 2 | 10.2391 | 7.2401 | 5.9115 | 5.1195 | 4.5791 |
| | 3 | 12.5818 | 8.8966 | 7.2641 | 6.2909 | 5.6267 |
| | 4 | 14.1945 | 10.0370 | 8.1952 | 7.0972 | 6.3478 |
| | 5 | 15.4067 | 10.8942 | 8.8951 | 7.7034 | 6.8901 |
| 5 | 1 | 5.6050 | 3.9633 | 3.2360 | 2.8025 | 2.5066 |
| | 2 | 9.5640 | 6.7628 | 5.5218 | 4.7820 | 4.2772 |
| | 3 | 11.9315 | 8.4368 | 6.8886 | 5.9657 | 5.3359 |
| | 4 | 13.5679 | 9.5939 | 7.8334 | 6.7839 | 6.0677 |
| | 5 | 14.7999 | 10.4651 | 8.5447 | 7.3910 | 6.6187 |
| 6 | 1 | 5.1166 | 3.6180 | 2.9541 | 2.5583 | 2.2882 |
| | 2 | 9.0539 | 6.4021 | 5.2273 | 4.5269 | 4.0490 |
| | 3 | 11.4374 | 8.0875 | 6.6034 | 5.7187 | 5.1150 |
| | 4 | 13.0908 | 9.2566 | 7.5580 | 6.5454 | 5.8544 |
| | 5 | 14.3376 | 10.1382 | 8.2778 | 7.1688 | 6.4119 |

**Table 8.7** (continued)

| m | n | $\lambda = 0.01, t = 10$ | $\lambda = 0.02, t = 10$ | $\lambda = 0.03, t = 10$ | $\lambda = 0.04, t = 10$ | $\lambda = 0.05, t = 10$ |
|---|---|---|---|---|---|---|
| 7 | 1 | 4.7371 | 3.3496 | 2.7349 | 2.3685 | 2.1185 |
| | 2 | 8.6490 | 6.1158 | 4.9935 | 4.3245 | 3.8679 |
| | 3 | 11.0432 | 7.8087 | 6.3758 | 5.5216 | 4.9387 |
| | 4 | 12.7095 | 8.9870 | 7.3378 | 6.3548 | 5.6839 |
| | 5 | 13.9677 | 9.8767 | 8.0643 | 6.9839 | 6.2465 |
| 8 | 1 | 4.4311 | 3.1333 | 2.5583 | 2.2156 | 1.9817 |
| | 2 | 8.3163 | 5.8805 | 4.8014 | 4.1581 | 3.7192 |
| | 3 | 10.7179 | 7.5787 | 6.1879 | 5.3589 | 4.7932 |
| | 4 | 12.3943 | 8.7641 | 7.1558 | 6.1971 | 5.5428 |
| | 5 | 13.6617 | 9.6602 | 7.8876 | 6.8308 | 6.1097 |
| 9 | 1 | 4.1777 | 2.9541 | 2.4120 | 2.0888 | 1.8683 |
| | 2 | 8.0359 | 5.6822 | 4.6395 | 4.0179 | 3.5937 |
| | 3 | 10.4422 | 7.3839 | 6.0289 | 5.2212 | 4.6610 |
| | 4 | 12.1269 | 8.5750 | 7.0015 | 6.0635 | 5.4233 |
| | 5 | 13.4020 | 9.4766 | 7.7376 | 6.7010 | 5.9935 |
| 10 | 1 | 3.9633 | 2.8025 | 2.2882 | 1.9817 | 1.7724 |
| | 2 | 7.7948 | 5.5117 | 4.5003 | 3.8974 | 3.4859 |
| | 3 | 10.2047 | 7.2158 | 5.8917 | 5.1023 | 4.5637 |
| | 4 | 11.8958 | 8.4116 | 6.8681 | 5.9479 | 5.3110 |
| | 5 | 13.1773 | 9.3178 | 7.6079 | 6.5887 | 5.8931 |

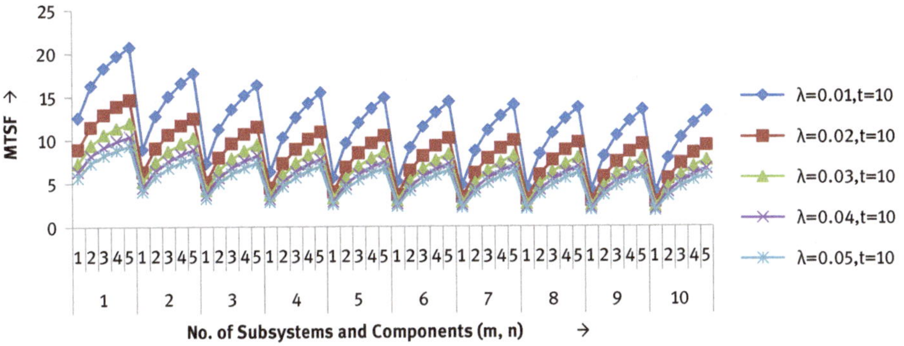


**Figure 8.8:** MTSF versus no. of subsystems (*m*) and components (*n*).

**Table 8.8:** Reliability versus no. of subsystems (*m*) and components (*n*).

| m | n | $\lambda=0.01, t=5$ | $\lambda=0.01, t=10$ | $\lambda=0.01, t=15$ | $\lambda=0.01, t=20$ | $\lambda=0.01, t=25$ |
|---|---|---|---|---|---|---|
| 1 | 1 | 0.8825 | 0.6065 | 0.3246 | 0.1353 | 0.0439 |
| | 2 | 0.9862 | 0.8452 | 0.5439 | 0.2523 | 0.0859 |
| | 3 | 0.9984 | 0.9391 | 0.6920 | 0.3535 | 0.1261 |
| | 4 | 0.9998 | 0.9760 | 0.7920 | 0.4410 | 0.1645 |
| | 5 | 0.99997 | 0.9906 | 0.8595 | 0.5167 | 0.2012 |
| 2 | 1 | 0.7788 | 0.3679 | 0.1054 | 0.0183 | 0.0019 |
| | 2 | 0.9726 | 0.7143 | 0.2958 | 0.0637 | 0.0074 |
| | 3 | 0.9967 | 0.8819 | 0.4788 | 0.1249 | 0.0159 |
| | 4 | 0.9996 | 0.9526 | 0.6272 | 0.1945 | 0.0271 |
| | 5 | 0.99995 | 0.9812 | 0.7388 | 0.2669 | 0.0405 |
| 3 | 1 | 0.6873 | 0.2231 | 0.0342 | 0.0025 | 0.00008 |
| | 2 | 0.9591 | 0.6037 | 0.1609 | 0.0161 | 0.0006 |
| | 3 | 0.9951 | 0.8281 | 0.3313 | 0.0442 | 0.0020 |
| | 4 | 0.9994 | 0.9298 | 0.4967 | 0.0858 | 0.0044 |
| | 5 | 0.99993 | 0.9720 | 0.6350 | 0.1379 | 0.0081 |

**Table 8.8** (continued)

| m | n | $\lambda=0.01, t=5$ | $\lambda=0.01, t=10$ | $\lambda=0.01, t=15$ | $\lambda=0.01, t=20$ | $\lambda=0.01, t=25$ |
|---|---|---|---|---|---|---|
| 4 | 1 | 0.6065 | 0.1353 | 0.0111 | 0.0003 | 0.0000037 |
| | 2 | 0.9459 | 0.5103 | 0.0875 | 0.0040 | 0.00005 |
| | 3 | 0.9935 | 0.7777 | 0.2293 | 0.0156 | 0.0002 |
| | 4 | 0.9992 | 0.9075 | 0.3934 | 0.0378 | 0.0007 |
| | 5 | 0.99991 | 0.9628 | 0.5458 | 0.0713 | 0.0016 |
| 5 | 1 | 0.5353 | 0.0821 | 0.0036 | 0.000045 | 0.00000016 |
| | 2 | 0.9328 | 0.4313 | 0.0476 | 0.0010 | 0.0000045 |
| | 3 | 0.9919 | 0.7303 | 0.1586 | 0.0055 | 0.000032 |
| | 4 | 0.9990 | 0.8858 | 0.3116 | 0.0167 | 0.0001 |
| | 5 | 0.99988 | 0.9537 | 0.4691 | 0.0368 | 0.0003 |
| 6 | 1 | 0.4724 | 0.0498 | 0.0012 | 0.0000061 | 0.000000007 |
| | 2 | 0.9110 | 0.3645 | 0.0259 | 0.00026 | 0.0000004 |
| | 3 | 0.9903 | 0.6858 | 0.1098 | 0.0019 | 0.000004 |
| | 4 | 0.9988 | 0.8645 | 0.2468 | 0.0073 | 0.00002 |
| | 5 | 0.99986 | 0.9447 | 0.4032 | 0.0190 | 0.00007 |
| 7 | 1 | 0.4169 | 0.0302 | 0.0004 | 0.00000083 | 0.0000000003 |
| | 2 | 0.9073 | 0.3081 | 0.0141 | 0.000065 | 0.000000035 |
| | 3 | 0.9887 | 0.6441 | 0.0760 | 0.0007 | 0.0000005 |
| | 4 | 0.9987 | 0.8438 | 0.1954 | 0.0032 | 0.0000033 |
| | 5 | 0.9998 | 0.9358 | 0.3465 | 0.0098 | 0.000013 |
| 8 | 1 | 0.3679 | 0.0183 | 0.0001 | 0.00000011 | 0.000000000014 |
| | 2 | 0.8947 | 0.2604 | 0.0076 | 0.000016 | 0.000000003 |
| | 3 | 0.9871 | 0.6048 | 0.0526 | 0.00024 | 0.00000006 |
| | 4 | 0.9985 | 0.8236 | 0.1548 | 0.0014 | 0.0000005 |
| | 5 | 0.99983 | 0.9270 | 0.2979 | 0.0051 | 0.000003 |

**Table 8.8** (continued)

| m | n | $\lambda = 0.01, t = 5$ | $\lambda = 0.01, t = 10$ | $\lambda = 0.01, t = 15$ | $\lambda = 0.01, t = 20$ | $\lambda = 0.01, t = 25$ |
|---|---|---|---|---|---|---|
| 9 | 1 | 0.3246 | 0.0111 | 0.00004 | 0.00000001 | 0.0000000000006 |
| | 2 | 0.8824 | 0.2201 | 0.0042 | 0.0000041 | 0.0000000003 |
| | 3 | 0.9855 | 0.5680 | 0.0364 | 0.000086 | 0.000000008 |
| | 4 | 0.9989 | 0.8038 | 0.1226 | 0.00063 | 0.00000009 |
| | 5 | 0.99980 | 0.9182 | 0.2560 | 0.0026 | 0.0000005 |
| 10 | 1 | 0.2865 | 0.0067 | 0.000013 | 0.000000002 | 0.00000000000003 |
| | 2 | 0.8702 | 0.1860 | 0.0023 | 0.0000011 | 0.00000000002 |
| | 3 | 0.9839 | 0.5334 | 0.0252 | 0.00003 | 0.000000001 |
| | 4 | 0.9981 | 0.7846 | 0.0971 | 0.0003 | 0.00000001 |
| | 5 | 0.9998 | 0.9096 | 0.2200 | 0.0013 | 0.0000001 |

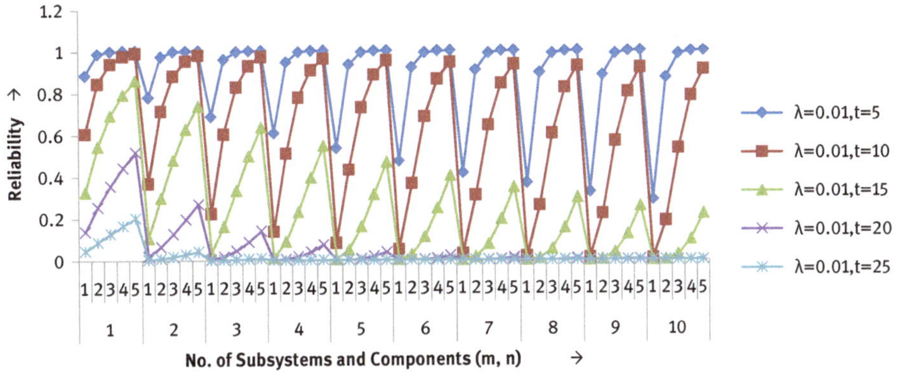


**Figure 8.9:** Reliability versus no. of subsystems (*m*) and components (*n*).

## 8.9 Exponential distribution

The exponential distribution is a special case of Weibull distribution with the shape parameter $\beta = 0$. the component reliability is given by

$$R_i(t) = e^{-\lambda_i t} \tag{8.10}$$

where $h_i(t) = \lambda_i$. Therefore, the system reliability is given by

$$R_s(t) = \prod_{j=1}^{m}\left[1 - \prod_{i=1}^{n}\left(1 - e^{-\lambda_i t}\right)\right] \tag{8.11}$$

where $1 - \prod_{i=1}^{n}(1 - R_i(t))$ is the reliability of a parallel system.

and

$$\text{MTSF} = \int_0^\infty R(t)dt = \int_0^\infty \prod_{j=1}^{m}\left[1 - \prod_{i=1}^{n}\left(1 - e^{-\lambda_i t}\right)\right]dt \tag{8.12}$$

For identical components, we have $\lambda_i = \lambda$

The system reliability is given by

$$R_s(\text{t}) = \left[1 - \left(1 - e^{-\lambda t}\right)^n\right]^m \tag{8.13}$$

And $\text{MTSF} = \int_0^\infty R_s(t)dt = \int_0^\infty \left[1 - \left(1 - e^{-\lambda t}\right)^n\right]^m \text{dt}$

Take $1 - e^{-\lambda t} = y$ so that $dt = \frac{dy}{\lambda(1-y)}$

Now $\text{MTSF} = \frac{1}{\lambda}\int_0^1 \frac{(1-y^n)^{m-1}}{(1-y)}dy = \frac{1}{\lambda}\int_0^1 (1-y^n)^{m-1}\frac{(1-y^n)}{(1-y)}dt$

Again take $y^n = z$ so that $y = z^{1/n}$ and

$dy = \frac{1}{nz^{\frac{n-1}{n}}}$

$$\text{MTSF} = \frac{1}{n\lambda}\sum_{i=0}^{n-1}\int_0^1 z^{\frac{i+1}{n}-1}(1-z)^{m-1}dz = \frac{1}{n\lambda}\sum_{i=0}^{n-1} B\left(\frac{i+1}{n}, m\right) = \frac{1}{n\lambda}\sum_{i=0}^{n-1}\frac{\Gamma\left(\frac{i+1}{n}\right)(m)}{\frac{i+1}{n}+m}$$

$$\text{MTSF} = \frac{1}{n\lambda}\sum_{i=0}^{n-1}\frac{\left(\frac{i+1}{n}-1\right)!(m-1)!}{\left(\frac{i+1}{n}+m-1\right)!} = \frac{(m-1)!}{n\lambda}\sum_{i=0}^{n-1}\frac{\left(\frac{i+1}{n}-1\right)!}{\left(\frac{i+1}{n}+m-1\right)!} \tag{8.14}$$

**Subcase**

Suppose system has only one subsystem, that is, $m = 1$, then

$$\text{MTSF} = \frac{1}{n\lambda}\sum_{i=0}^{n-1}\frac{1}{i+1/n} = \frac{1}{\lambda}\sum_{i=0}^{n-1}\frac{1}{i+1} = \frac{1}{\lambda}\left[1 + \frac{1}{2} + \frac{1}{3} + \cdots + \frac{1}{n}\right] = \frac{1}{\lambda}\sum_{i=1}^{n}\frac{1}{i}$$

The result obtained for MTSF in this case is same as the MTSF of the parallel system of "$n$" components.

## 8.10 Reliability measures for arbitrary values of the parameters

Reliability and MTSF of the system have been obtained for arbitrary values of the parameters associated with number of subsystems ($m$), number of components ($n$), failure rate ($\lambda$) and operating time of the component ($t$). The results are shown numerically in the tables 8.9 to 8.11 and graphically in the figures 8.10 to 8.12.

**Table 8.9:** Reliability versus no. of subsystems ($m$) and components ($n$).

| m | n | $\lambda = 0.01, t = 10$ | $\lambda = 0.02, t = 10$ | $\lambda = 0.03, t = 10$ | $\lambda = 0.04, t = 10$ | $\lambda = 0.05, t = 10$ |
|---|---|---|---|---|---|---|
| 1 | 1 | 0.9048 | 0.8187 | 0.7408 | 0.6703 | 0.6065 |
| | 2 | 0.9909 | 0.9671 | 0.9328 | 0.8913 | 0.8452 |
| | 3 | 0.9991 | 0.9940 | 0.9826 | 0.9642 | 0.9391 |
| | 4 | 0.9999 | 0.9989 | 0.9955 | 0.9882 | 0.9760 |
| | 5 | 0.999992 | 0.9998 | 0.9988 | 0.9961 | 0.9906 |
| 2 | 1 | 0.8187 | 0.6703 | 0.5488 | 0.4493 | 0.3678 |
| | 2 | 0.9820 | 0.9354 | 0.8702 | 0.7944 | 0.7143 |
| | 3 | 0.9983 | 0.9881 | 0.9655 | 0.9296 | 0.8819 |
| | 4 | 0.9998 | 0.9978 | 0.9910 | 0.9765 | 0.9526 |
| | 5 | 0.99998 | 0.9996 | 0.9977 | 0.9922 | 0.9812 |
| 3 | 1 | 0.7408 | 0.5488 | 0.4066 | 0.3012 | 0.2231 |
| | 2 | 0.9731 | 0.9046 | 0.8117 | 0.7081 | 0.6037 |
| | 3 | 0.9974 | 0.9822 | 0.9487 | 0.8963 | 0.8281 |
| | 4 | 0.9997 | 0.9968 | 0.9865 | 0.9650 | 0.9298 |
| | 5 | 0.99998 | 0.9994 | 0.9965 | 0.9884 | 0.9720 |
| 4 | 1 | 0.6703 | 0.4493 | 0.3012 | 0.2019 | 0.1353 |
| | 2 | 0.9643 | 0.8749 | 0.7572 | 0.6311 | 0.5103 |
| | 3 | 0.9965 | 0.9764 | 0.9321 | 0.8642 | 0.7777 |
| | 4 | 0.9997 | 0.9956 | 0.9821 | 0.9536 | 0.9075 |
| | 5 | 0.99997 | 0.9992 | 0.9953 | 0.9845 | 0.9628 |

**Table 8.9** (continued)

| m | n | $\lambda = 0.01, t = 10$ | $\lambda = 0.02, t = 10$ | $\lambda = 0.03, t = 10$ | $\lambda = 0.04, t = 10$ | $\lambda = 0.05, t = 10$ |
|---|---|---|---|---|---|---|
| 5 | 1 | 0.6065 | 0.3679 | 0.2231 | 0.1353 | 0.0821 |
| | 2 | 0.9555 | 0.8461 | 0.7063 | 0.5625 | 0.4313 |
| | 3 | 0.9957 | 0.9706 | 0.9159 | 0.8332 | 0.7303 |
| | 4 | 0.9996 | 0.9946 | 0.9776 | 0.9423 | 0.8858 |
| | 5 | 0.99996 | 0.9990 | 0.9941 | 0.9807 | 0.9537 |
| 6 | 1 | 0.5488 | 0.3012 | 0.1653 | 0.0907 | 0.0498 |
| | 2 | 0.9469 | 0.8183 | 0.6589 | 0.5014 | 0.3645 |
| | 3 | 0.9948 | 0.9648 | 0.8999 | 0.8034 | 0.6858 |
| | 4 | 0.9995 | 0.9935 | 0.9732 | 0.9312 | 0.8645 |
| | 5 | 0.99995 | 0.9988 | 0.9930 | 0.9768 | 0.9447 |
| 7 | 1 | 0.4966 | 0.2466 | 0.1224 | 0.0608 | 0.0302 |
| | 2 | 0.9383 | 0.7915 | 0.6146 | 0.4469 | 0.3081 |
| | 3 | 0.9940 | 0.9590 | 0.8843 | 0.7746 | 0.6441 |
| | 4 | 0.9994 | 0.9925 | 0.9688 | 0.9202 | 0.8438 |
| | 5 | 0.99994 | 0.9986 | 0.9918 | 0.9730 | 0.9358 |
| 8 | 1 | 0.4493 | 0.2019 | 0.0907 | 0.0408 | 0.0183 |
| | 2 | 0.9298 | 0.7654 | 0.5733 | 0.3983 | 0.2604 |
| | 3 | 0.9931 | 0.9533 | 0.8689 | 0.7468 | 0.6048 |
| | 4 | 0.9993 | 0.9914 | 0.9645 | 0.9093 | 0.8236 |
| | 5 | 0.99994 | 0.9984 | 0.9907 | 0.9693 | 0.9270 |
| 9 | 1 | 0.4066 | 0.1653 | 0.0672 | 0.0273 | 0.0111 |
| | 2 | 0.9214 | 0.7403 | 0.5348 | 0.3550 | 0.2201 |
| | 3 | 0.9923 | 0.9476 | 0.8538 | 0.7201 | 0.5680 |
| | 4 | 0.9993 | 0.9903 | 0.9601 | 0.8986 | 0.8038 |
| | 5 | 0.99993 | 0.9982 | 0.9895 | 0.9655 | 0.9182 |

**Table 8.9** (continued)

| m | n | $\lambda=0.01, t=10$ | $\lambda=0.02, t=10$ | $\lambda=0.03, t=10$ | $\lambda=0.04, t=10$ | $\lambda=0.05, t=10$ |
|---|---|---|---|---|---|---|
| 10 | 1 | 0.3679 | 0.1353 | 0.0498 | 0.0183 | 0.0067 |
| | 2 | 0.9130 | 0.7160 | 0.4989 | 0.3164 | 0.1860 |
| | 3 | 0.9914 | 0.9420 | 0.8389 | 0.6943 | 0.5334 |
| | 4 | 0.9992 | 0.9892 | 0.9558 | 0.8879 | 0.7846 |
| | 5 | 0.99992 | 0.9980 | 0.9884 | 0.9617 | 0.9096 |

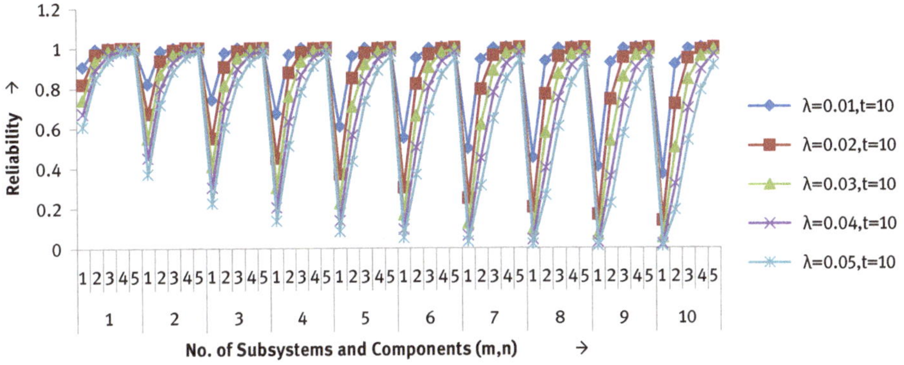


**Figure 8.10:** Reliability versus no. of subsystems (*m*) and components (*n*).

**Table 8.10:** MTSF versus no. of subsystems (*m*) and components (*n*).

| m | n | $\lambda=0.01, t=10$ | $\lambda=0.02, t=10$ | $\lambda=0.03, t=10$ | $\lambda=0.04, t=10$ | $\lambda=0.05, t=10$ |
|---|---|---|---|---|---|---|
| 1 | 1 | 100 | 50 | 33.3333 | 25 | 20.0000 |
| | 2 | 150 | 75 | 50 | 37.5 | 30 |
| | 3 | 183.3333 | 91.6667 | 61.1111 | 45.8333 | 36.6667 |
| | 4 | 208.3333 | 104.1667 | 69.4444 | 52.0833 | 41.6667 |
| | 5 | 228.3333 | 114.1667 | 76.1111 | 57.0833 | 45.6667 |
| 2 | 1 | 50 | 25 | 16.6667 | 12.5 | 10 |
| | 2 | 91.6667 | 45.8333 | 30.5556 | 22.9167 | 18.3333 |
| | 3 | 121.6667 | 60.8333 | 40.5556 | 30.4167 | 24.3333 |
| | 4 | 144.881 | 72.4405 | 48.2936 | 36.2202 | 28.9762 |
| | 5 | 163.7698 | 81.8849 | 54.5899 | 40.9425 | 32.7540 |

**Table 8.10** (continued)

| m | n | $\lambda = 0.01, t = 10$ | $\lambda = 0.02, t = 10$ | $\lambda = 0.03, t = 10$ | $\lambda = 0.04, t = 10$ | $\lambda = 0.05, t = 10$ |
|---|---|---|---|---|---|---|
| 3 | 1 | 33.3333 | 16.6667 | 11.1111 | 8.3333 | 6.6667 |
| | 2 | 70 | 35 | 23.3333 | 17.5 | 14 |
| | 3 | 97.8968 | 48.9484 | 32.6323 | 24.4742 | 19.5794 |
| | 4 | 119.9639 | 59.9820 | 39.9880 | 29.9910 | 23.9928 |
| | 5 | 138.1324 | 69.0662 | 46.0441 | 34.5331 | 27.6265 |
| 4 | 1 | 25 | 12.5 | 8.3333 | 6.25 | 5 |
| | 2 | 58.2143 | 29.1071 | 19.4048 | 14.5536 | 11.6429 |
| | 3 | 84.5996 | 42.2998 | 28.1999 | 21.1499 | 16.9199 |
| | 4 | 105.8304 | 52.9152 | 35.2768 | 26.4576 | 21.1661 |
| | 5 | 123.47 | 61.7350 | 41.1567 | 30.8675 | 24.694 |
| 5 | 1 | 20.0000 | 10 | 6.6667 | 5 | 4 |
| | 2 | 50.6349 | 25.3175 | 16.8783 | 12.6587 | 10.1270 |
| | 3 | 75.8525 | 37.9262 | 25.2842 | 18.9631 | 15.1705 |
| | 4 | 96.4297 | 48.2148 | 32.1432 | 24.1074 | 19.2859 |
| | 5 | 113.6534 | 56.8267 | 37.8845 | 28.4133 | 22.7307 |
| 6 | 1 | 16.6667 | 8.3333 | 5.5556 | 4.1667 | 3.3333 |
| | 2 | 45.2742 | 22.6371 | 15.0914 | 11.3185 | 9.0548 |
| | 3 | 69.5471 | 34.7735 | 23.1824 | 17.3868 | 13.9094 |
| | 4 | 89.5901 | 44.7951 | 29.8634 | 22.3975 | 17.9180 |
| | 5 | 106.4723 | 53.2362 | 35.4908 | 26.6181 | 21.2945 |
| 7 | 1 | 14.2857 | 7.1428 | 4.7619 | 3.5714 | 2.8571 |
| | 2 | 41.2421 | 20.6210 | 13.7474 | 10.3105 | 8.2484 |
| | 3 | 64.7259 | 32.3630 | 21.5753 | 16.1815 | 12.9452 |
| | 4 | 84.3189 | 42.1594 | 28.1063 | 21.0797 | 16.8638 |
| | 5 | 100.912 | 50.456 | 33.6373 | 25.228 | 20.1824 |

**Table 8.10** (continued)

| m | n | $\lambda = 0.01, t = 10$ | $\lambda = 0.02, t = 10$ | $\lambda = 0.03, t = 10$ | $\lambda = 0.04, t = 10$ | $\lambda = 0.05, t = 10$ |
|---|---|---|---|---|---|---|
| 8 | 1 | 12.5 | 6.25 | 4.1667 | 3.125 | 2.5 |
| | 2 | 38.0759 | 19.0380 | 12.6920 | 9.5190 | 7.6152 |
| | 3 | 60.8850 | 30.4425 | 20.2950 | 15.2212 | 12.1770 |
| | 4 | 80.0900 | 40.045 | 26.6967 | 20.0225 | 16.018 |
| | 5 | 96.4330 | 48.2165 | 32.1443 | 24.1082 | 19.2866 |
| 9 | 1 | 11.1111 | 5.55556 | 3.7037 | 2.7778 | 2.2222 |
| | 2 | 35.5094 | 17.7547 | 11.8365 | 8.8773 | 7.1019 |
| | 3 | 57.7309 | 28.8654 | 19.2436 | 14.4327 | 11.5462 |
| | 4 | 76.5956 | 38.2978 | 25.5319 | 19.1489 | 15.3191 |
| | 5 | 92.7185 | 46.5085 | 30.9062 | 23.1796 | 18.5437 |
| 10 | 1 | 10 | 5 | 3.3333 | 2.5 | 2 |
| | 2 | 33.3773 | 16.6887 | 11.1258 | 8.3443 | 6.6755 |
| | 3 | 55.0798 | 27.5399 | 18.3599 | 13.7610 | 11.0160 |
| | 4 | 73.6421 | 36.8210 | 24.5474 | 18.4105 | 14.7284 |
| | 5 | 89.5686 | 44.7843 | 29.8562 | 22.3921 | 17.9137 |

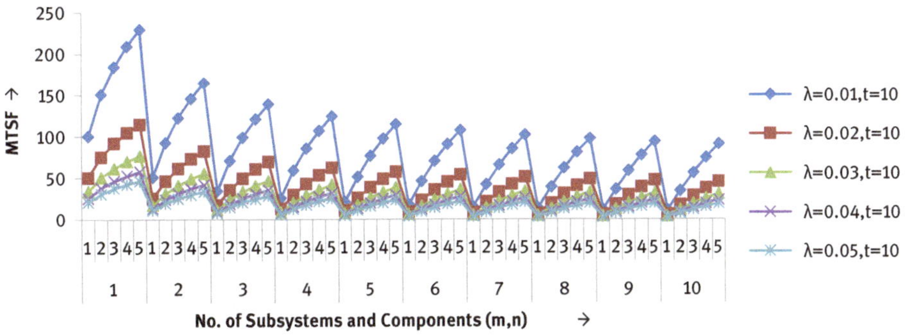


**Figure 8.11:** MTSF versus no. of components (*n*) and subsystems (*m*).

**Table 8.11:** Reliability versus no. of subsystems (*m*) and components (*n*).

| m | n | $\lambda = 0.01, t = 5$ | $\lambda = 0.01, t = 10$ | $\lambda = 0.01, t = 15$ | $\lambda = 0.01, t = 20$ | $\lambda = 0.01, t = 25$ |
|---|---|---|---|---|---|---|
| 1 | 1 | 0.9512 | 0.9048 | 0.8607 | 0.8187 | 0.7788 |
| | 2 | 0.9976 | 0.9909 | 0.9806 | 0.9671 | 0.9511 |
| | 3 | 0.99988 | 0.9991 | 0.9973 | 0.9940 | 0.9892 |
| | 4 | 0.999994 | 0.99992 | 0.9996 | 0.9989 | 0.9976 |
| | 5 | 0.9999997 | 0.999992 | 0.99995 | 0.9998 | 0.9995 |
| 2 | 1 | 0.9048 | 0.8187 | 0.7408 | 0.6703 | 0.6065 |
| | 2 | 0.9952 | 0.9820 | 0.9616 | 0.9354 | 0.9045 |
| | 3 | 0.9998 | 0.9983 | 0.9946 | 0.9881 | 0.9785 |
| | 4 | 0.999989 | 0.9998 | 0.9992 | 0.9978 | 0.9952 |
| | 5 | 0.999999 | 0.99998 | 0.9998 | 0.9996 | 0.9989 |
| 3 | 1 | 0.8607 | 0.7408 | 0.6376 | 0.5488 | 0.4724 |
| | 2 | 0.9929 | 0.9731 | 0.9429 | 0.9046 | 0.8603 |
| | 3 | 0.9996 | 0.9974 | 0.9919 | 0.9822 | 0.9679 |
| | 4 | 0.99998 | 0.9997 | 0.9989 | 0.9968 | 0.9928 |
| | 5 | 0.999999 | 0.99998 | 0.9998 | 0.9994 | 0.9984 |
| 4 | 1 | 0.8187 | 0.6703 | 0.5488 | 0.4493 | 0.3679 |
| | 2 | 0.9905 | 0.9643 | 0.9246 | 0.8749 | 0.8182 |
| | 3 | 0.9995 | 0.9965 | 0.9892 | 0.9765 | 0.9574 |
| | 4 | 0.99998 | 0.9997 | 0.9985 | 0.9957 | 0.9904 |
| | 5 | 0.999999 | 0.99997 | 0.9998 | 0.9992 | 0.9979 |
| 5 | 1 | 0.7788 | 0.6065 | 0.4724 | 0.3679 | 0.2865 |
| | 2 | 0.9882 | 0.9555 | 0.9067 | 0.8461 | 0.7781 |
| | 3 | 0.9994 | 0.9957 | 0.9866 | 0.9706 | 0.9470 |
| | 4 | 0.99997 | 0.9996 | 0.9981 | 0.9946 | 0.9880 |
| | 5 | 0.999999 | 0.99996 | 0.9997 | 0.9990 | 0.9973 |

**Table 8.11** (continued)

| m | n | $\lambda = 0.01, t = 5$ | $\lambda = 0.01, t = 10$ | $\lambda = 0.01, t = 15$ | $\lambda = 0.01, t = 20$ | $\lambda = 0.01, t = 25$ |
|---|---|---|---|---|---|---|
| 6 | 1 | 0.7408 | 0.5488 | 0.4066 | 0.3012 | 0.2231 |
| | 2 | 0.9858 | 0.9469 | 0.8891 | 0.8183 | 0.7401 |
| | 3 | 0.9993 | 0.9948 | 0.9839 | 0.9648 | 0.9368 |
| | 4 | 0.99997 | 0.9995 | 0.9977 | 0.9935 | 0.9857 |
| | 5 | 0.999998 | 0.99995 | 0.9997 | 0.9988 | 0.9968 |
| 7 | 1 | 0.7047 | 0.4966 | 0.3499 | 0.2466 | 0.1738 |
| | 2 | 0.9835 | 0.9383 | 0.8718 | 0.7915 | 0.7039 |
| | 3 | 0.9992 | 0.9940 | 0.9812 | 0.9590 | 0.9266 |
| | 4 | 0.99996 | 0.9994 | 0.9974 | 0.9925 | 0.9834 |
| | 5 | 0.999998 | 0.99994 | 0.9996 | 0.9986 | 0.9963 |
| 8 | 1 | 0.6703 | 0.4493 | 0.3012 | 0.2019 | 0.1353 |
| | 2 | 0.9811 | 0.9298 | 0.8549 | 0.7654 | 0.6694 |
| | 3 | 0.9991 | 0.9931 | 0.9786 | 0.9533 | 0.9166 |
| | 4 | 0.99995 | 0.9993 | 0.9970 | 0.9914 | 0.9810 |
| | 5 | 0.999998 | 0.99994 | 0.9996 | 0.9984 | 0.9958 |
| 9 | 1 | 0.6376 | 0.4066 | 0.2592 | 0.1653 | 0.1054 |
| | 2 | 0.9788 | 0.9214 | 0.8383 | 0.7403 | 0.6367 |
| | 3 | 0.9989 | 0.9923 | 0.9759 | 0.9476 | 0.9067 |
| | 4 | 0.99995 | 0.9993 | 0.9966 | 0.9903 | 0.9786 |
| | 5 | 0.999998 | 0.99993 | 0.9995 | 0.9982 | 0.9952 |
| 10 | 1 | 0.6065 | 0.3679 | 0.2231 | 0.1353 | 0.0821 |
| | 2 | 0.9765 | 0.9130 | 0.8221 | 0.7160 | 0.6055 |
| | 3 | 0.9988 | 0.9914 | 0.9733 | 0.9420 | 0.8969 |
| | 4 | 0.99994 | 0.9992 | 0.9962 | 0.9892 | 0.9763 |
| | 5 | 0.999997 | 0.99992 | 0.9995 | 0.9980 | 0.9947 |

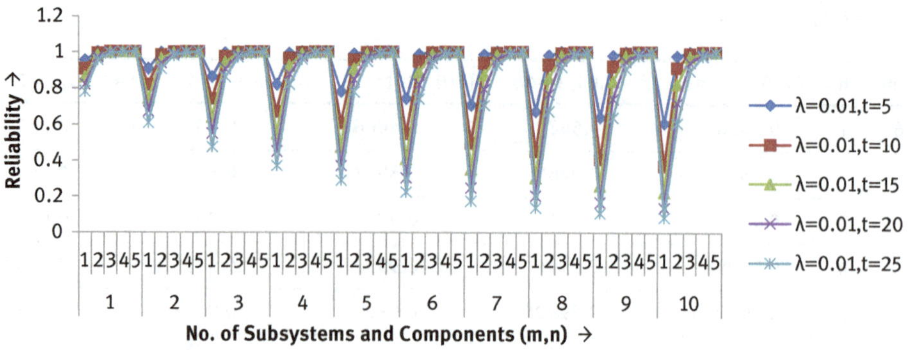


**Figure 8.12:** Reliability versus no. of subsystems (*m*) and components (*n*).

# 8.11 Discussion and conclusion

Here, we discussed a series–parallel system consisting "*m*" subsystems connected in series and each having "*n*" components connected in parallel. The effect of number of components, failure rates of the components, shape parameter, and operating time of the components on reliability and MTSF has been examined by considering Weibull failure laws. It is observed that reliability and MTSF keeps on increasing with the increase of components (connected in parallel) in subsystems while their value declines with the increase of subsystems (connected in series), failure rate, and operating time of the components. For a special case of Weibull failure laws, that is, Rayleigh and exponential failure laws, the values of reliability, and MTSF have also been observed. But, the reliability and MTSF in case of exponential failure laws is more as compared to Weibull and Rayleigh failure laws. Further, reliability of a series–parallel system goes on decreasing with the increase of operating time irrespective of the distributions related to failure time of components and subsystems. For Weibull failure laws, the reliability and MTSF keeps on decreasing with the increase in the shape parameter.

Hence, finally we conclude that the performance of a series–parallel system can be improved by increasing the number of components rather than to increase the number of subsystems. The results are shown numerically and graphically in the respective tables and figures. It is interesting to note that reliability of this system is more than that of series system while a parallel system has more reliability.

# References

[1] Barlow, R.E. & Prochan, F. (1975). Statistical Theory of Reliability and Life Testing, Holt, Rinehart & Winston, Inc, New York.
[2] E. Balagurusamy (1984). Reliability Engineering, Tata McGraw Hill Publishing Co. Ltd., India.
[3] L. S. Srinath (1985). Concept in Reliability Engineering, Affiliated East-West Press (P) Ltd.
[4] Chauhan, S.K. and Malik, S.C. (2016). Reliability evaluation of a series-parallel and parallel-series systems for arbitrary values of the parameters, International Journal of Statistics and Reliability Engineering, 3(1), 10–19.
[5] Elsayed, A. (2012). Reliability Engineering, Wiley Series in Systems Engineering and Management.
[6] Dao, C.D., Zuo, M.J. and Pandey, M. (2014). Selective maintenance for multi-state series-parallel systems under economic dependence, Reliability Engineering System Safety, 121, 240–249.
[7] M.A. El-Damcese (2009). Reliability equivalence factors of sereis-parallel system in Weibull distribution, International Mathematical Forum, 4(1), 941–951.
[8] Sarhan, A.M. (2002). Reliability equivalence with a basic series-parallel system, Applied mathematics and computation, 132(1), 115–133.
[9] Misra, K.B. (1972). Reliability optimization of a series-parallel system, IEE Transactions on Reliability, R-21(4), 230–238.
[10] Ei-Heweihi, E., Proschan, F. and Sethuraman, J. (1986). Optimal allocation of components in parallel-series and series-parallel system, Journal of Applied Probability, 23, 770–777.
[11] Gopalan, M.N. (1975). Availability and reliability of a series-parallel system with a single repair facility, IEEE Transactions on Reliability, R–24(3), 219–220.

# Index



https://doi.org/10.1515/9783110619058-009

# De Gruyter Series on the Applications of Mathematics in Engineering and Information Sciences

**Already published in the series**

**Volume 3: Computational Intelligence. Theoretical Advances and Advanced Applications**
Dinesh C. S. Bisht, Mangey Ram (Eds.)
ISBN 978-3-11-065524-7, e-ISBN (PDF) 978-3-11-067135-3
e-ISBN (EPUB) 978-3-11-066833-9

**Volume 2: Supply Chain Sustainability. Modeling and Innovative Research Frameworks**
Sachin Kumar Mangla, Mangey Ram (Eds.)
ISBN 978-3-11-062556-1, e-ISBN (PDF) 978-3-11-062859-3,
e-ISBN (EPUB) 978-3-11-062568-4

**Volume 1: Soft Computing. Techniques in Engineering Sciences**
Mangey Ram, Suraj B. Singh (Eds.)
ISBN 978-3-11-062560-8, e-ISBN (PDF) 978-3-11-062861-6,
e-ISBN (EPUB) 978-3-11-062571-4

# De Gruyter Series on the Applications of Mathematics in Engineering and Information Sciences

[illegible]